Plastic Pipe Systems

Plastic Pipe Systems
Failure Investigation and Diagnosis

Mehdi Farshad
Swiss Federal Laboratories for Materials
Testing and Research
EMPA, Switzerland

2006

ELSEVIER

AMSTERDAM • BOSTON • HEIDELBERG • LONDON • NEW YORK • OXFORD
PARIS • SAN DIEGO • SAN FRANCISCO • SINGAPORE • SYDNEY • TOKYO

Butterworth-Heinemann is an imprint of Elsevier
Linacre House, Jordan Hill, Oxford OX2 8DP, UK
30 Corporate Drive, Suite 400, Burlington, MA 01803, USA

First edition 2006
Reprinted 2007

Notice
No responsibility is assumed by the publisher for any injury and/or damage to persons
or property as a matter of products liability, negligence or otherwise, or from any use
or operation of any methods, products, instructions or ideas contained in the material
herein. Because of rapid advances in the medical sciences, in particular, independent
verification of diagnoses and drug dosages should be made

British Library Cataloguing in Publication Data
A catalogue record for this book is available from the British Library

Library of Congress Cataloging-in-Publication Data
A catalog record for this book is available from the Library of Congress

ISBN: 978-1-856-17496-1

For information on all Butterworth-Heinemann publications
visit our website at books.elsevier.com

Printed and bound in *China*

07 08 09 10 10 9 8 7 6 5 4 3 2

Working together to grow
libraries in developing countries

www.elsevier.com | www.bookaid.org | www.sabre.org

ELSEVIER   BOOK AID International   Sabre Foundation

Contents

8   Clogging of the pipe system   166

9   The knowledge base of an expert system   172
   for failure diagnosis

About the author

Prof. Dr. Mehdi Farshad has academic degrees in civil engineering and engineering mechanics including B.S. from Tehran University, M.S. and C.E. from Columbia University, and Ph.D. in from Stanford University in USA. From 1971 to 1991 he was engineering professor at the University of Shiraz. During this period, he spent 2 years as guest professor at the University of Toronto, Canada. Since 1991, Prof. Farshad has been a Senior Research Scientist at EMPA (Swiss Federal Laboratories for Materials Testing and Research) in Switzerland and Titular Professor at ETH (Swiss Federal Institute of Technology) in Zürich, where he has been teaching courses in several branches of engineering sciences and conducting applied research and technical consulting activities. His areas of specialties include polymeric and composite materials, structural engineering, shell structures, stability, finite element simulations, biomechanics, and pipeline engineering. Prof. Farshad is the author of about 200 research papers and 30 books, including "Design and analysis of shell structures" and "Stability of Structures". He has lead several international research programs and has been involved in conception, design, and realization of numerous professional engineering projects and technical expertise, failure investigations, and structural rehabilitation. He is the member of several international standardization committees, an external evaluator for SAS (Swiss Federal Accreditation) organization, a member of the Chamber of Swiss Experts, founding member of academy of sciences, and winner of several awards and honors. Prof. Farshad is the developer of the pipeline structural analysis and design program ADAP and the founder of an EMPA spin-off the *Farshad Technical Consulting* enterprise in Switzerland (www.farshad.ch).

Preface

Pipe systems are one of the most reliable and safest means of transfer of mater and energy. They are, in fact, the *lifelines* of communities. New materials such as plastic products and composites have enhanced the domain of application of material systems in pipelines. Plastic pipes have salient features such as low weight, ease of connection, and corrosion resistance. Like any other installations, however, statistical cases of malfunction and failure may occasionally occur in piping systems. An objective failure investigation of such incidents can prove to be very instructive for health monitoring, safety, maintenance, retrofitting, and life cycle management of pipelines. Hence, the knowledge gained from numerous failure investigations together with a sound scientific and engineering basis should become available to manufacturers, planers, users, and engineers dealing with piping systems. Due to ageing of the existing pipelines, this topic may gain more importance in the coming years. Therefore, the need for systematic investigation procedure, proper diagnosis, health monitoring, and decisions as well as the rehabilitation methods is expected to increase in the future.

In this book, a reference guideline dealing with the failure analysis of various pipes and in particular plastic and composite pipes is presented. The main motivations for composition of this book have been: (1) need of a wide class of users, planers, and engineers for a simple pipe diagnosis system and (2) availability of a fair amount of experience and data on the failure analyses in the part of the author and his colleagues. This book is the result of many years of research, teaching, and practice as well as the experience in various failure investigations. In the first chapter of this book, an overview of polymeric materials, their properties, and plastic pipes are presented. The second chapter deals with a systematic procedure for failure investigation of plastic pipes. In Chapters 3–9 various potential modes of failure in plastic pipes are discussed. These chapters include basic features and underlying mechanisms of failure modes including crazing, cracking, fracture, buckling and large deformations, changes of color and dimension, local damages, delamination, corrosion, and clogging of plastic piping systems. Each of the above-mentioned chapters includes several case studies related to actual failure cases occurred in practice. The last chapter of this book presents a knowledge base for the plastic pipe failure investigations. The skeleton of this knowledge base can be used to create an expert system for a systematic pipe failure diagnosis. A glossary of plastics pipes is also enclosed.

This book has a condensed and reference-type configuration enriched with numerous tabular and graphic and photographic representations. It also contains numerous examples of actual failure cases occurred in pipelines, photographs of the failed pipe system, and a condensed description and diagnosis of the failure event. These case studies can be used as sample procedural guidelines for other cases. This book can be used as a reference manual by a relatively wide spectrum of users including technical personal of communities responsible for pipelines, engineers and planers, pipe manufacturers, pipeline authorities, and standardization organizations.

I would like to thank the EMPA colleagues and organizations, who directly or indirectly contributed to the collection of data, the failure investigation reports, and the useful discussions and collaborations in joint failure investigation and also to Dr. Peter Richner from EMPA for his support of this publication. I would like to express my gratitude to my wife, Gowhartaj, for her several decades of support, love, devotion, and care.

M. Farshad
Zürich, Switzerland

1
Plastic pipe systems

1.1 Polymeric materials

1.1.1 Definition and origin of polymeric materials

A *polymer* (natural or synthetic) is a very large molecule made up of many smaller units joined together to create a long chain. Synthetic polymers are materials constituted of long molecular chains (macromolecules) and organic connections obtained through processing of natural products or through *synthesis* of primary materials from oil, gas, or coal. The smallest "building block" of a polymer is called a *monomer*. If all the monomers are chemically the same, then that polymer is called a homopolymer (Fig. 1.1).

Monomers generally contain carbon and hydrogen with, sometimes, other elements such as oxygen, nitrogen, chlorine, or fluorine. The most common example of a synthetic homopolymer is polyethylene; other common materials are polypropylene, polystyrene, and polyvinyl chloride, more commonly known as "PVC".

Synthetic polymers are divided into three distinct groups: thermoplastics, thermosetting, and elastomeric polymers. The *thermoplastics* are those which, once formed, can be heated and reformed over and over again. *Thermosetting polymers* cannot be reformed or remolded. Thermosetting plastics differ from thermoplastics chemically in that heating

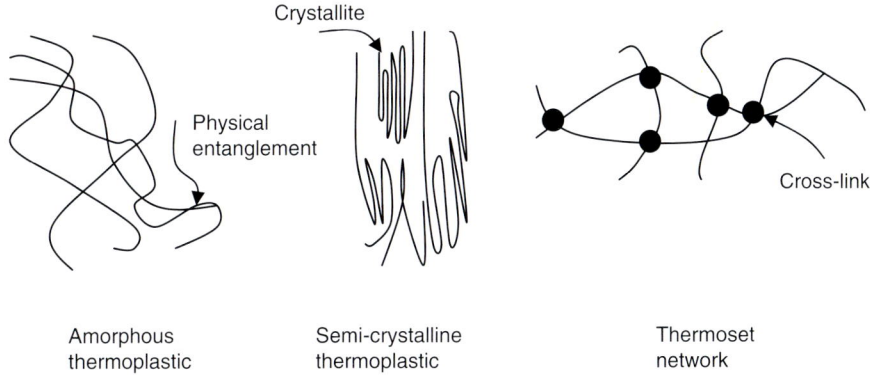

Fig. 1.1. Schematics of various types of polymer molecular structures.

Table 1.1 Classes of synthetic polymeric materials.

Polymer types	General feature	Examples
Plastomers (thermoplasts)	"Spaghetti" structure Strong softening with temperature; reversible hardening Medium strength Large deformation capacity	PVC PE PP PC PMMA
Thermosetting plastics, duromers, duroplasts	Strong cross-link No softening with temperature High strength Small strain capacity	GF-UP, GF-EP, GF-VI, PUR
Elastomers	Loose spatial cross-links Flow at high temperature Low strength Large strain capacity	Natural rubber Synthetic rubbers PUR PVC-P
TPE	Rubber elastic Medium strength Large strain capacity	TPU

PE: polyethylene; PP: polypropylene; PC: polycarbonate; PMMA: Polymethylmethacrylate; GF-UP: glass fiber with unsaturated polyester matrix; GF-EP: glass fiber with epoxy matrix; GF-VI; glass fiber with vinylester matrix; PUR: polyurethane; PVC-P: plasticized PVC; TPE: thermoplastic elastomers; TPU: thermoplastic polyurethane.

the former introduces a three-dimensional network to the long chains so that they are no longer able to flow freely past one another like they can in the case of the thermoplastics. *Rubbers*, more properly called *elastomers*, are the third group of polymeric materials. Elastomers are polymers with relatively weak cross-linking. Hence, they are capable of large recoverable strains up to several hundred percent. Table 1.1 summarizes the classes of synthetic materials.

1.1.2 Polymers and plastics

Distinction is usually made between *polymers* and *plastics*. To enhance the properties of polymers for the desired applications, some substances are added to polymers; these are called *additives*. A plastic is by definition the polymer material, which has some degree of additives in it. Table 1.2 presents a number of additives and their function in plastic materials.

1.1.3 Physical properties of plastics

The behavior of plastic pipes are strongly influenced by the plastic materials and polymeric components used in production of the pipes. Plastic materials have certain salient features that distinguish them from other materials such as metals and ceramics. The

Table 1.2 Additives and their function in plastics.

Additive	Function
Stabilizers	Heat stabilization, prevention of depolymerization
Antioxidants	Prevention of oxidative degradation
Modifiers	To change some properties (e.g., impact resistance)
UV protection agents	UV protection, prevention of photo-oxidative degradation
Colorants	Absorption of incident light
Flame retardants	Fire protection, inhibition of pyrolysis
Antistatics	Reduction of charge building in polymers
Lubricants	Ease of production, reduction of internal friction
Plasticizers	Reduction of viscosity and stiffness
Coupling agents	Improvement of surface properties
Biocides	Prevention of microbiological degradation of the polymer
Reinforcing agents	Increase of strength and stiffness
Blowing agents	Vaporization for foaming
Special additives (e.g., magnetic powders)	To produce certain functions in the polymer (e.g., electric or magnetic conductivity)

UV: ultraviolet.

main feature of plastics and polymer composites is that their physical, mechanical, thermal, and chemical properties are strongly time and temperature dependent. Some of the important properties of polymeric materials and products include the tensile properties, the softening and melting behavior, the viscoelastic behavior, the temperature behavior, and the so-called hygro-thermal response.

The tensile properties of polymers and polymer composites are obtained from the direct or indirect tensile tests on the polymeric material and plastic pipe samples. The short-term and the long-term tensile properties are temperature dependent and are also different from each other. The major parameters identifying the tensile properties are tensile modulus, yield point, tensile strength, rupture stress, and maximum strain. Fig. 1.2 schematically shows a typical tensile behavior of thermoplastic materials.

The stiffness of a polymeric material is also affected by the temperature. In general, a reduction of stiffness is to be expected at higher temperatures. For thermoplastic materials, the reduction of stiffness at elevated temperatures is quite remarkable. For example, at the elevated temperature of about 60°C, the tensile modulus of polyethylene can be reduced to about 50% of the modulus at the room temperature. Fig. 1.3 shows the reduction of tensile modulus of some thermoplastic materials at elevated temperatures.

An important characteristic index of polymers representing the change of behavior of polymers due to temperature is called *glass tension temperature* (Tg). The glass transition temperature signifies the temperature at which the flow behavior of a plastic material undergoes a qualitative transition. Namely, at temperatures lower than Tg, the amorphous polymers are in a glassy and hard state; above Tg, the amorphous polymers

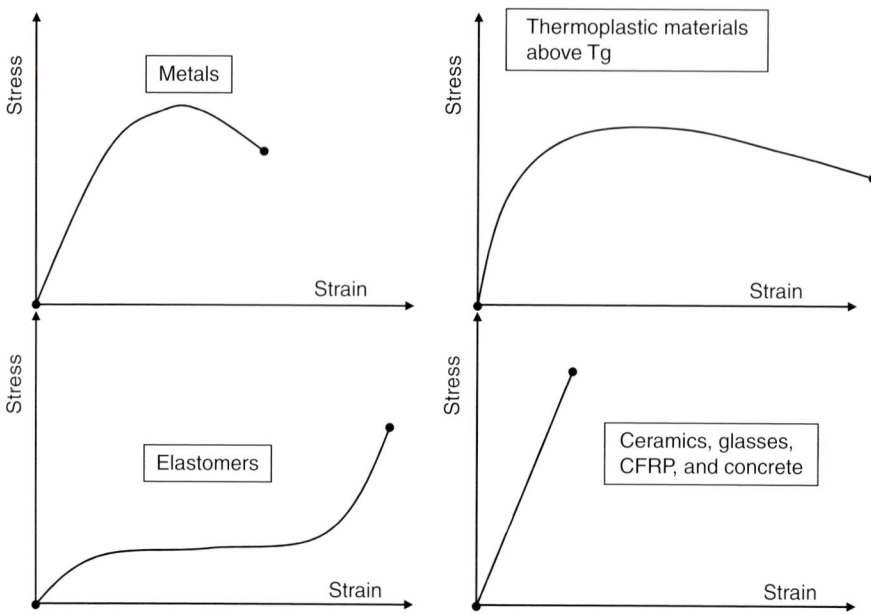

Fig. 1.2. General qualitative tensile properties of various materials.

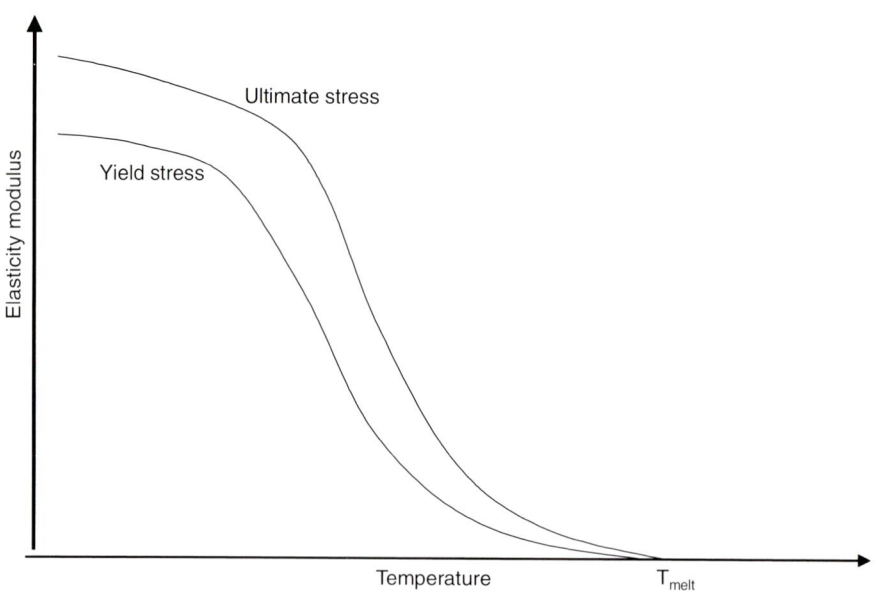

Fig. 1.3. Qualitative temperature dependence of the tensile properties of a
thermoplastic material.

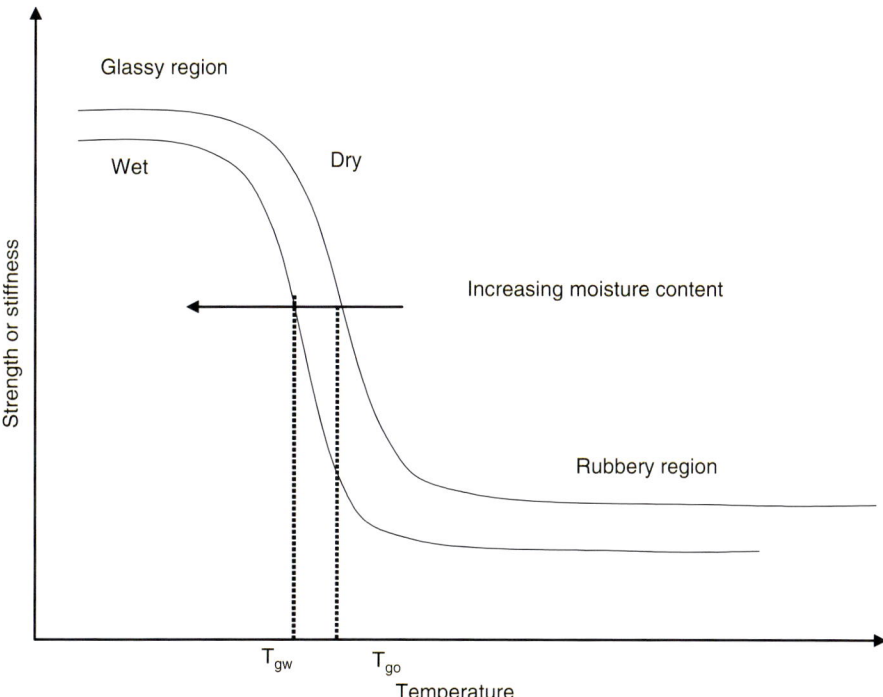

Fig. 1.4. Definition of glass transition temperature.

are in a rubbery or viscous state. In the part-crystalline polymers, the temperature behavior is mostly determined by the amorphous part. The mechanism for this transition is the increase of the movement of the chain molecules with the increase of temperature. Fig. 1.4 schematically shows the general temperature-dependent behavior of polymers and the definition of the glass transition temperature.

Polymeric materials are generally viscoelastic and are capable of creep behavior. Under constant applied load, a polymeric material shows an instantaneous elastic behavior, but undergoes further time-dependent creep deformation (Fig. 1.5). Upon removal of the load, the elastic part of the deformation is recovered and a certain amount of deformation remains as the permanent plastic deformation. The reverse situation is called *stress relaxation*. Fig. 1.6 schematically depicts the general creep behavior of polymeric materials.

One of the important properties of fiber-reinforced polymers is known as the hygro-thermal effect. As the name implies, the hygro-thermal properties are the result of interaction of moisture and thermal factors. In the wet condition, fiber-reinforced plastics generally undergo dimensional change, structural change, and change of properties. The external temperature can strongly affect the moisture response of fiber-reinforced polymers. The interaction of these two factors is nonlinear and the resulting property, that is, the hygro-thermal effects can strongly influence the mechanical properties and the service life of fiber-reinforced pipes. Fig. 1.7 schematically shows the hygro-thermal properties of fiber-reinforced plastics.

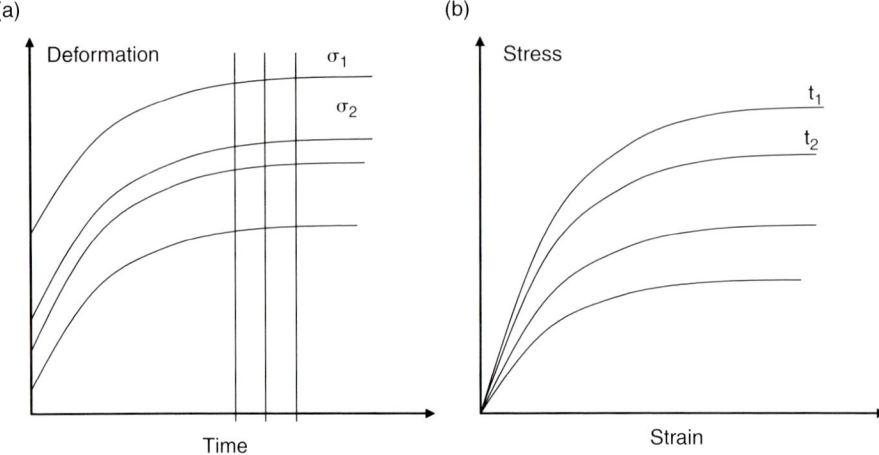

Fig. 1.5. Time-dependent property of polymeric materials. (a) Creep deformation at various constant stress levels and (b) isochrones (stress versus strain at specified time in creep loading).

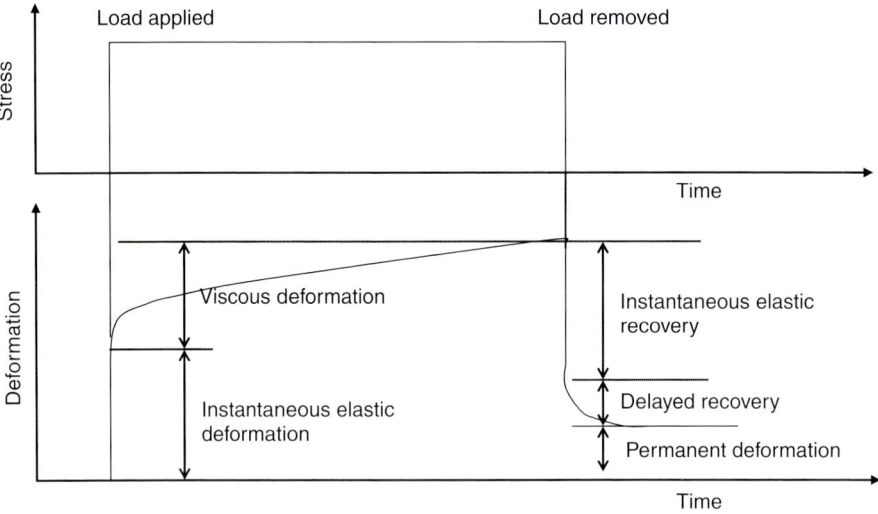

Fig. 1.6. Creep behavior of plastic materials.

The material properties of various plastics not fixed numbers, but normally have a range of variation, which is influenced by the material composition, production of the sample, external medium, time, and temperature. Hence fixed material property values, as normally mentioned, for example, in the case of metals, cannot be cited for plastic materials. Nevertheless, some typical values representing the *order of magnitude* of properties can be cited. These values can specially be useful as benchmarks for the plastic pipe investigations. Table 1.3 shows the typical values of some properties for number of plastic materials.

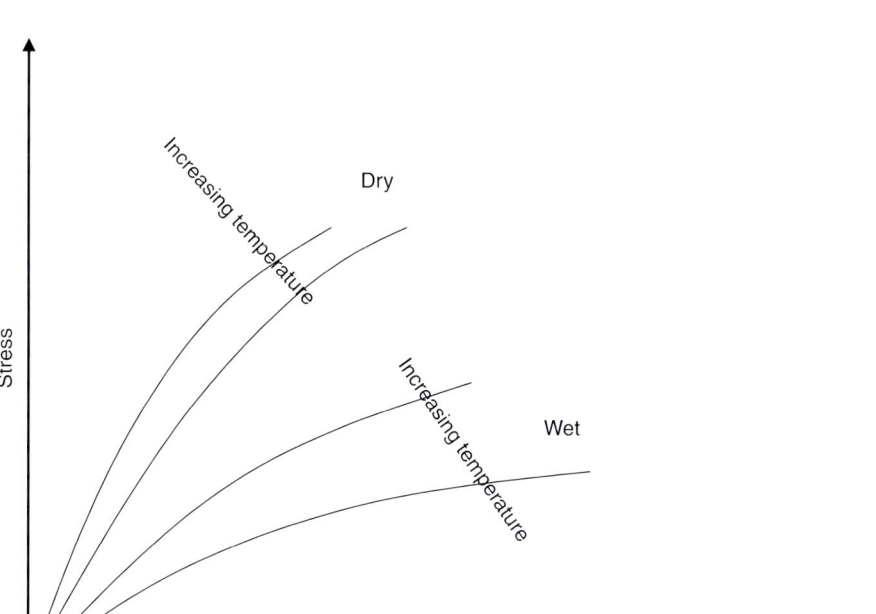

Fig. 1.7. Effect of temperature and moisture on the tensile properties of polymers.

1.2   Plastic pipes

Piping systems for gas and water distribution, sewer, and drainage systems, cable protection, communication, and industrial installations constitute the lifelines of various industries and communities. Many thousand kilometer long existing pipelines around the world made of metallic, concrete, polymeric, and composite materials perform their vital function with various degrees of efficiency, but generally with high degree of safety. A number of the existing piping networks are, however, locally or globally aged or are prone to potential damage and failure. Some piping systems may have even reached the limit of their service lifetime and may be at the stage of potential failure or may need retrofitting or even replacement. Statistically, each year a number of failures occur in the pipelines; some of which cause material damages and even endanger life. The main question regarding an existing pipeline deals with reliability and remaining service life of the system. This issue becomes critical when one or more cases of malfunction or failures have happened in a certain piping system.

In the case of plastic pipes, the above-mentioned questions and comments find additional material-specific dimensions. Plastics piping systems have proved quite reliable for gas, water, and drainage systems as well as several other applications. Plastic pipes are light, easy to connect, resistant against corrosion, flexible, and easy to handle. The plastic materials, however, have their own salient features, which should be taken into consideration in all issues dealing with the safety, service life, failure event, and the retrofitting strategies.

Table 1.3 Short-term mechanical and thermal typical properties of some selected plastics at T = 23 °C.

Plastic material	Short designation	Density (kg/m³)	Tensile strength (N/mm²)	Tearing strain (%)	Tensile modulus (N/mm²)	Coefficient of linear thermal expansion (°K × 10⁻⁶)
Polyethylene, high density	PE-HD	0.94–0.96	18–35	100–1000	700–1400	200
Cross-linked high-density polyethylene	PE-HDX	0.955	25	15	1000	100–120
Polypropylene	PP	0.90–0.907	21–37	20–800	1100–1300	150
Mineral-reinforced polypropylene	PP-QD	11.5		2800		
Polybutene-1	PB	0.905–0.920	30–38	250–280	250–350	150
Polyvinylchloride	PVC-U	1.38–1.55	50–75	10.0–50	1000–3500	70–80
Polytetrafluorethylene	PTFE	2.15–2.20	25–36	350–550	410	100
Polymethylmethacrylate	PMMA	1.17–1.20	50–77	<10	2700–3200	70
Polyacetal	POM	1.41–1.42	62–70	25–70	2800–3200	90–110
Polyamide 6	PA 6	1.13	70–85	200–300	1400	80
Polyamide 66	PA 66	1.14	77–84	150–300	2000	80
Polyamide 12	PA 12	1.02	56–65	300	1600	150
Polycarbonate	PC	1.2	56–67	100–130	2100–2400	60–70
Polysulfone	PSU	1.24	50–100	25–30	2600–2750	54
Polyvinylidene fluoride	PVDF	1.78	53.4	80	2400	128
Polyetheretherketone	PEEK	1.32	90	50	3600	47
Epoxy resin	EP	1.9	30–40	4	21,500	11.0–35

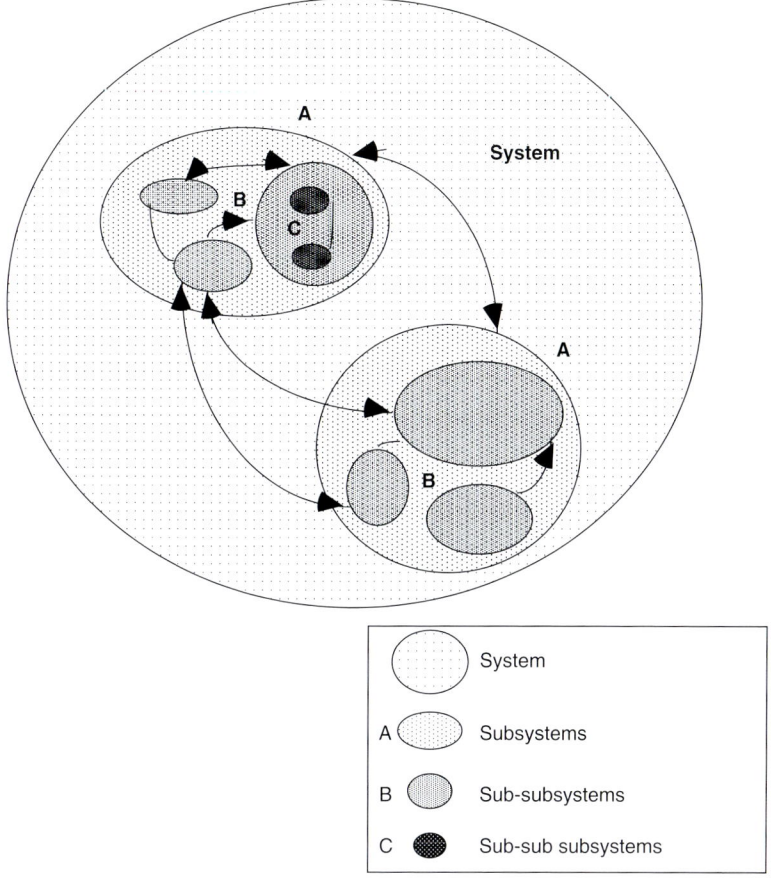

Fig. 1.8. Systems and subsystems.

1.3   Pipes as systems

A *system* can be defined as a collection of *related* and/or *interacting elements*. The elements of a system may themselves consist of systems and may constitute their own subsystems, sub-subsystems, and so on (Fig. 1.8). A *subsystem* is an element or a functional component of a larger system; it fulfils the requirements of a systemic entity which has been integrated into the larger system and, as a constituting part, plays a role in the existence and behavior of the larger system.

A system is always identified in relation with its *environment*. The *environment* of a system can be defined as the set of elements, parameters, variables, attributes, and phenomena that are not considered to be part of the defined system, but they may exert some influence on the system; conversely, they may be affected by the system. Fig. 1.9 represents this definition of the environment. A system may be a *subsystem* of its environment, that is, it may be an element of a super-system.

From the systems point of view, pipelines, like any other technical installation, are not only an assemblage of elements, but are *structural systems* interacting with their

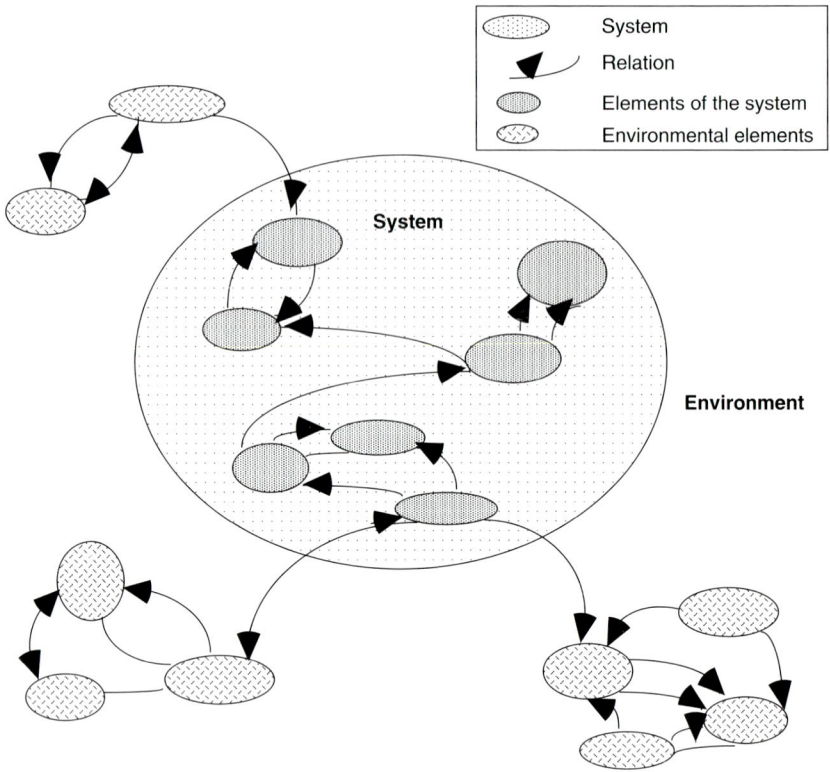

Fig. 1.9. System and its environment.

environment. For instance, the partial settlement of the soil around the buried pipe will change the pressure applied to the pipe and the stiffness of the pipe in relation with the stiffness of soil will determine the distribution of the soil pressure.

The systemic character of pipeline installations becomes more prominent in plastic pipes. Plastic pipes are made of polymers, which demonstrate time-dependent properties. Hence, ageing of polymeric materials brings the pipe systems closer to the level of biological entities. This emphasis gains great importance in the study of the service life of pipe systems and the limiting conditions leading to the failure of pipe systems.

1.4   Classification of plastic pipes

Plastic pipes may be classified according to several methods. Two of the mostly applied methods of classifications are:

(1) Classification on the basis on geometrical parameters and resistance to internal pressure.
(2) Classification according to ring stiffness.

The first type of classification is mostly used for pressure thermoplastic pipes. The second type of classification is usually applied to drainage pipes and composite pipes such as glass fiber-reinforced plastics (GFRP) pipes. In this section, a brief account of these two classifications is presented.

The notation used for the definitions are:

D_m: Average diameter of the pipe
D_e: External diameter of the pipe
DN: Nominal diameter of the pipe (for thermoplastic pipes, equal to the external diameter)
e: Pipe wall thickness
δD_v: Change in the pipe ring diameter
k: A numerical factor for nonlinear behavior of ring und diametrical force
E: Elastic modulus of the pipe in the circumferential direction
v: Poisson's ratio
E_i: Short-term elastic modulus of the pipe in the circumferential direction
E_t: Long-term elastic modulus of the pipe in the circumferential direction
I: Second moment of area of the pipe wall section:

$$I = e^3/12(1 - v^2)$$

SR: Pipe ring stiffness:

$$SR = 8EI/D_m^3$$

F: Diametrical force on the pipe ring

1.4.1   Classification on the basis on geometrical parameters and resistance to internal pressure

- Pipe series, S, is defined as the ratio of average pipe radius to the wall thickness:

$$S = (d_e - e)/2e = D_m/2e$$

- Standard diameter ratio, SDR, is defined as the pipe external diameter to the wall thickness:

$$SDR = D_e/e$$

- Nominal pressure, PN, is defined as the resistance to internal hydrostatic pressure at room temperature after 50 years.
- Minimum required strength, MRS, is defined as the rounded value of the pipe strength to internal hydrostatic pressure at room temperature after 50 years multiplied by 10. An example of this classification is: polyethylene, high density (PE-HD), PE 100, $D_e = 315$ mm, $e = 28.6$ mm, $S = 5$, SDR = 11, PN = 16 bar.

1.4.2   Classification according to ring stiffness

Ring stiffness according to diametrical force tests on the pipe rings:

$$S = kF/\delta D_v$$

Calculated ring stiffness:

$$S = EI/D_m^3 = SR/8$$

Short-term ring stiffness:

$$Si = E_iI/D_m^3$$

Long-term ring stiffness:

$$St = E_tI/D_m^3$$

An example of this classification is a GFRP pipe with nominal diameter DN = 500 mm, and wall thickness e = 10.7 mm and nominal pressure PN = 10 bar. The calculated ring stiffness of the pipe is SR = 0.04 N/mm². The stiffness classification of the pipe based on this stiffness value would be: SN = SR/8 = 5000 N/m².

1.5   Requirements for the piping systems

Depending on the material, function, application, and the service environment of a specific pipeline project, a number of technical requirements are defined. Fig. 1.10 presents various categories of requirements on the pipe systems. In relation with the failure analysis, one may define the failure event as the phenomenon in which one or several of such requirements are no longer fulfilled (more precise definition in Chapter 2).

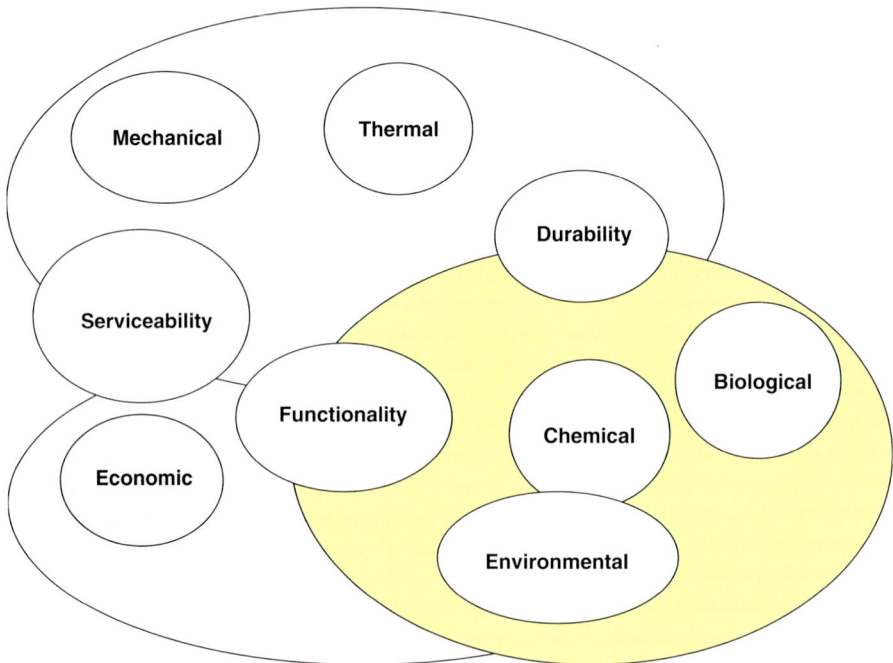

Fig. 1.10. Requirements on piping systems.

The general technical requirements on the plastic pipe systems may be listed as follows:

(1) Strength requirements related to limit loads, stresses, and strains
(2) Ductility
(3) Fracture toughness
(4) Fatigue resistance
(5) Impact resistance
(6) Stiffness (limited displacements and deformations)
(7) Creep response, long-term behavior
(8) Durability
(9) Stability
(10) Serviceability
(11) Fire resistance
(12) Functionality
(13) Abrasion resistance
(14) Corrosion resistance
(15) Thermal stability
(16) Chemical resistance
(17) Biological criteria
(18) Environmental requirements
(19) Recycling features

1.6   Environmental effects

Plastic pipeline may be subjected to various types of environmental conditions. The active environment may include mechanical loads, thermal effects, radiation, chemical influences, and biological factors. These conditions may individually or collectively cause a failure mode in the pipeline. The interaction of these factors is not a linear event, but may occur in a nonlinear fashion and may cause nonlinear magnifications and new qualitative consequences. The hygro-thermal effects in composite materials are an example of nonlinear interaction of the thermal and wet environment. The environmental stress cracking (ESC) (discussed in Chapter 2) is another example in which the nonlinear interaction of mechanical stresses with chemical factors may lead to the cracking modes of failure.

In defining and assessing the environmental effects one should regard a pipeline as a *system*. A pipeline is composed of various elements including pipes, connections, fixes points, and connected machinery which have interactions with each other and with the external environment. The piping system itself may also have inherent stresses, deformations, and strains. For example, the residual stresses in the pipe may influence the behavior of the elements as well as the whole piping system.

The so-called *environmental effects* are external actions imparted to the pipeline as a material system. The major environmental factors on the pipeline are as follows:

(1) Mechanical actions:
  ● Internal pressure
  ● Negative pressure (internal vacuum)

- External pressure
- Earth pressure
- Traffic loads
- Overburden
- Water pressure
- Surface loads
- Length effects (axial tension, compression, and bending)
- Point loads
- Impact forces
- Forces due to change of direction and fix points
- Frost
- Dynamic effects (vibrations, fatigue)
- Hydrodynamic forces
- Buoyancy forces
- Residual stresses, initial stress
- Forces during production
- Actions during storage and transport
- Load cases during installation
- Landslides, fault movements
- Earthquake
- Soil liquefaction with lateral spreading
- Washouts
- Unforeseeable loads (bomb damage, impacts, new environmental actions)
- Forces arising from interaction with foundation, supports, and other installations
- Partial settlement, earth subsidence
- Effects due to repairs, checks, intervention actions

(2) Thermal effects:
- Thermal effects, temperature gradients
- Longitudinal forces caused by temperature changes
- Thermal ageing of the material
- Temperature effects due to hydration in the concrete embedded pipes

(3) Chemical effects:
- Water
- Alkaline medium
- Acids
- Solvents
- Oil
- Oxygen

(4) Biological factors:
- Microbes and bacteria
- Animals

(5) Long-term effects (ageing factors)

The major categories of the environmental factors may be identified as shown in Fig. 1.11. Fig. 1.12 summarizes the material types of plastic pipes, the various pipe systems, the environmental effects, and the potential failure events in plastic pipe systems.

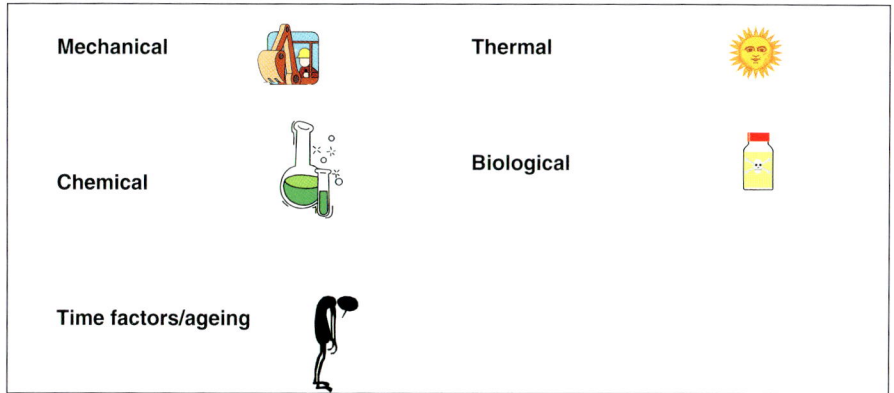

Fig. 1.11. Major environmental effects on piping systems.

1.7   Life cycle of plastic pipes

1.7.1   *Service life engineering*

The life cycle of material systems consists of their conception, development, service life, ageing, disintegration, and recycling of the material system. The important aspects in the life cycle engineering are:

(1)   Ageing and potential failure mechanisms
(2)   Durability of the system
(3)   Long-term behavior of materials and the system as a whole
(4)   Reliability analysis
(5)   Sustainability
(6)   Recycling

Interaction between the technical system and the non-technical environment the so-called service life engineering (SLE) is a module of the life cycle engineering. It deals with the interval of development and usage of the material systems. In the case of plastic pipes, the SLE includes the research and development activities related to the material and system properties and functions. The activities in this field include failure analyses, development of new materials and experimental methods for determination of the long-term properties, and development of extrapolation methods to predict the service life of polymer-based systems up to the periods of 50 and 100 years.

Many events related to the pipelines are statistical in nature. The failure probability of pipelines, like any other installation, increases as they age. Even new piping systems may experience an "early life" failure due to various reasons. The SLE takes into account the statistical events and the probability of their occurrence. Hence many engineers, communities, and authorities are faced with the statistical question regarding the health of an existing pipeline, its service life, and the means of failure prevention and rehabilitation. Obviously, any decision regarding the rehabilitation or renewal of these ageing systems would need a systematic study of the state of health, potential failure events, and their causes. In this connection, failure statistics and the results of failure case

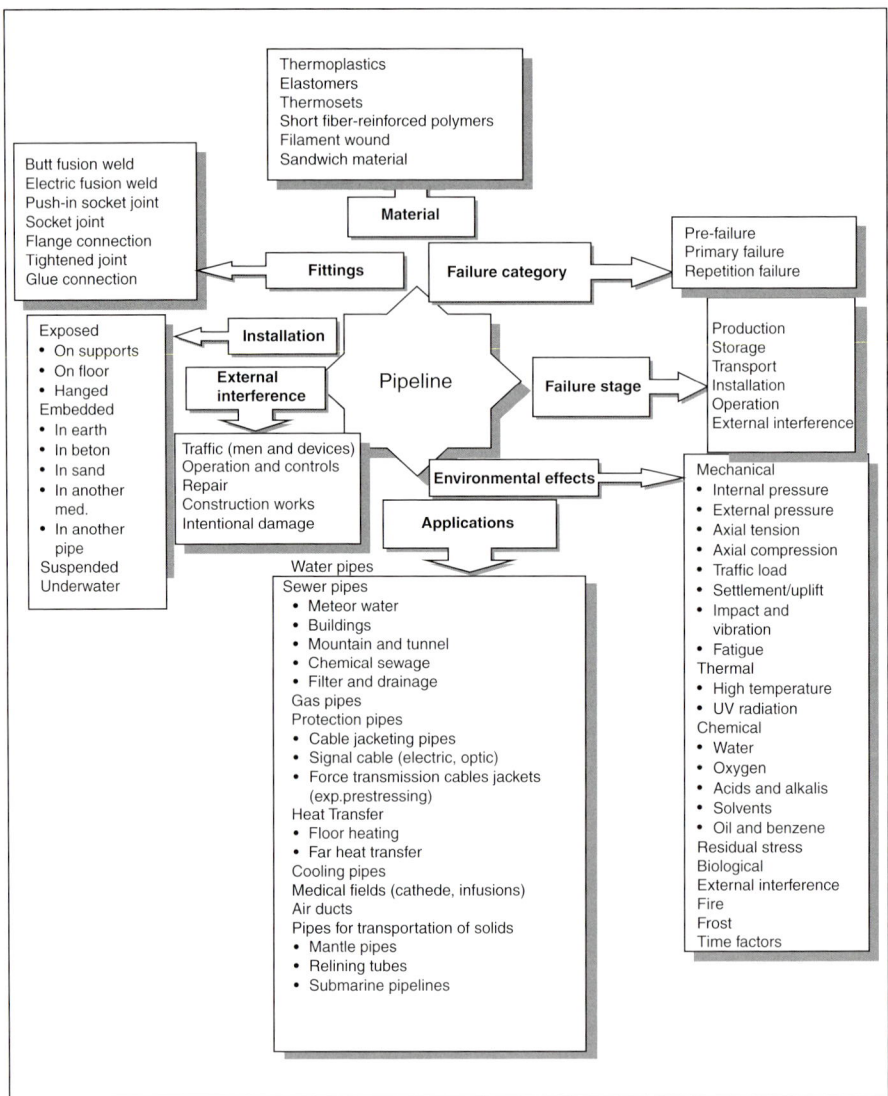

Fig. 1.12. Plastic pipes, their general features, applications, and potential failure categories.

studies provide valuable information on the damage mechanisms and measures to prevent the unexpected failure of pipelines.

1.7.2   The bathtub theory

Technical systems and man-made products may be viewed as having a certain life cycle. Biologically stated, technical systems may be viewed as being some sort of *living*

Table 1.4 Failure rate according to the bathtub theory.

Phase	Features of failure rate	Cases
Early failures	High failure probability, but decreasing failure rate	Material defects, quality changes in production, installation, application errors, dimensioning error, and operational errors
Random failures	Relatively constant failure rate	Random failures and spontaneous failures of otherwise sound systems (e.g., buckling failures, interventions)
Wear-out failures	Increasing failure rate toward higher failure probability	Ageing, deterioration, fatigue

systems. Many technical systems manifest the variety of the features, which are normally attributed to the living organisms: They come to being, they grow and interact with their environment, they age, and they die. The external influences such as modifications, repairs, changes, and demolitions may affect the life cycle of the technical systems and engineered products.

The organismic analogy of technical systems and engineered products with the biological and living systems can be assessed by a common behavior, which is normally expressed as the *bathtub theory*. The events dealing with the behavior of all time-dependent systems including plastic pipe systems may be explained by the *bathtub theory*. The bathtub theory presents the probability of failure of all time-dependent systems as function of the time parameter. According to this theory, the failure probabilities are relatively high at the initial stage of the systems life; the probability remains relatively constant for a certain time period, and then rises again at the final stages of the life of the system. The name "bathtub" comes from the similarity of the probability curve with a longitudinal section of a "bathtub," hence the name the *Bathtub theory*. Table 1.4 summarizes the three stages of service life of the mechanical systems.

Fig. 1.13 is a graphic representation of the bathtub theory. According to the bathtub theory, the lifespan of the things consists of three *risk stages*:

(1) The initial stage is the period of birth and initial development. In the case of technical systems, this stage is the initial period of utilization and operation of a new system; for example, a dam, a bridge, an airplane, a ship, and products of everyday usage. With the passage of time, the existence of the newly born organism or newly constructed man-made system is established. In Fig. 1.13 this stage is represented by a curve of descending trend of the failure probability.

(2) The stage of steady-state existence is a period of the lifespan in which the system operates normally or develops in a steady-state rhythm with relatively low risk of failure. In Fig. 1.13 this stage is represented by a curve of relatively constant value of the failure probability.

(3) The stage of ageing and decay signifies the last period of the lifecycle of a system. As represented in Fig. 1.13, the probability of failure increases in time until the death of the system with the probability of one is reached.

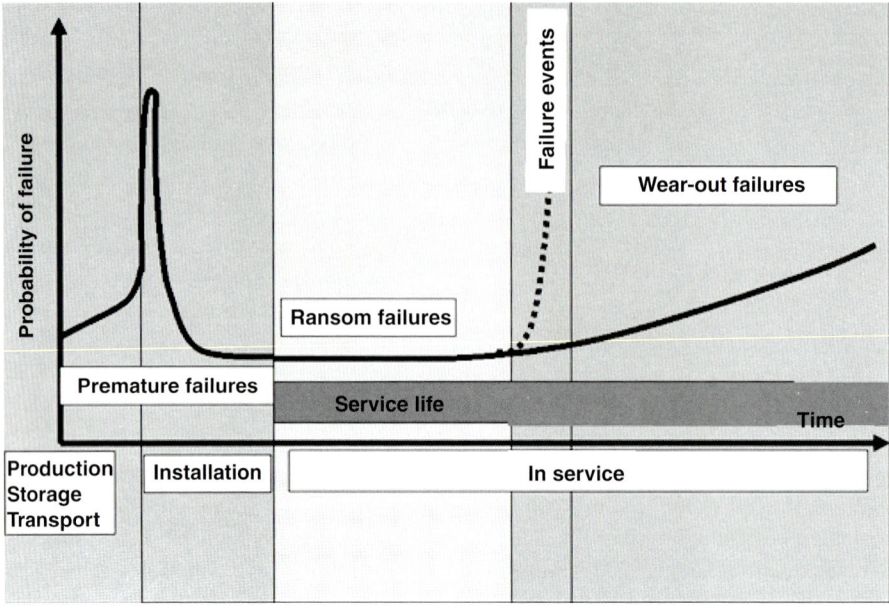

Fig. 1.13. Geometrical representation of the bathtub theory as related to the failure probability of piping system.

With imposition of the external influences the probability of failure can be reduced or increased. Environmental hygiene and medical interventions, in the case of living systems, and repairs and changing of damaged parts, in the case of man-made products are means by which the probability of failure may be reduced. The result of such actions is the lower curves in Fig. 1.13.

1.8   Initial stresses in plastic pipes

Initial stresses are internal stresses, which may exist in a body prior to exposition to an external environment. These stresses can be generated during production of the object; hence they are also referred to as the *residual stresses* or *frozen-in stresses*. Initial stresses may play a significant role in the behavior of the objects in mechanical, thermal, and chemical environments. In chemical environments, the interaction of initial stresses with chemical agents can lead to the phenomenon known as the *environmental stress corrosion* or *environmental stress cracking*. This phenomenon is of importance in failure diagnosis of some material systems.

Initial stresses may be generated in plastic pipes during production phases. Thermoplastic pipes are extruded from an extruder the outcome of which is a relatively soft and hot tubular object. The pipe is then normally cooled with water from outside and during lowering of temperature a hard pipe shape is produced. During the cooling process, a thermal gradient is produced in the pipe wall thickness; this temperature difference can generate initial stresses in circumferential direction as well as in the length direction.

Fig. 1.14. Thin rings from a 22 mm thick PE-HD pipe (diameter 225 mm, SDR 11, pipe series 5) after 1000 h. Two top photos are consecutive pictures of the rings from the outer side of the pipe wall thickness. The two bottom figures are consecutive photos of the inner side of the pipe wall thickness. The thickness of each ring was about 10% of the pipe wall thickness. The split edges of the outer rings overlap each other by an amount of about 127 mm, while those of the inner rings move away from each other by an amount of approximately 5.3 mm.

Normally, through an external cooling a compression hoop stress at the outer side and a tensile hoop stress in the inner side of the pipe wall is produced. The stress variation in the pipe wall thickness is generally nonlinear.

There are several methods to determine the initial stresses in plastic pipes. One of such methods is to cut thin rings from the pipe wall thickness and then make a slit in each ring in the longitudinal direction. The initial hoop stress causes the slits to move away from each other or tend to come together. Fig. 1.14 shows this method applied to polyethylene pipe ring section. The magnitude of initial stresses and the distribution of residual stress field are highly dependent on the material and the processing technique. For some thermoplastic pipes, the initial stresses may amount to about 10% of tensile strength of the material.

1.9   Ageing of plastic pipes

1.9.1   Long-term behavior of plastic pipes

Consideration of the long-time behavior of polymer pipes and polymer-based composite pipes is crucial for many practical purposes. The importance of this matter arises from the fact that the behavior of polymeric materials is highly time and temperature dependent.

A number of phenomena can be related to the long-term behavior of polymers and plastic pipes. The long-term response of plastic materials manifests itself in change of some chemical, physical, and mechanical properties. In macroscopic level, change in strength and stiffness are two indices, which can be associated with the long-term behavior. Table 1.5 gives an overview of the phenomena, which may lead to change of strength or stiffness in plastic material.

For pressure pipes under internal pressure, the main issue is the long-term behavior at the service temperature and the related safety factors. For other applications, the internal hydrostatic pressure tests provide a sound basis for characterization of the long-term material and pipe behavior. Generally speaking, at high temperatures the internal pressure resistance of polymer pipes is appreciably smaller than lower service temperatures.

Polymeric pipes with sustained internal pressure are normally designed to have 50 years of safe operation. Considering the time dependency in behavior of such pipes, it is crucial to have an estimation of pressure resistance of polymeric pipelines during long time of operation. Obviously, a long-time testing of sample of the produced pipes is improbable. An appropriate method of estimating the long-time behavior of these pipes would be to perform creep rupture tests up to certain period of time and, then, by means of a proper extrapolation, to obtain an estimate of there long-time behavior.

The present guidelines for the design of plastic pipes use the short-term and the long-term stresses in the pipe and apply the safety factor to the long-term hoop stress under hydrostatic pressure. According to the existing procedure, the internal hydrostatic pressure creep rupture tests and their extrapolation to 50 years of the pipe service life is considered as one of the main design and evaluation criteria. The creep rupture tests are normally carried out up to about 2 years at various temperatures.

In order to determine the long-term behavior and to estimate the service lifetime of plastic pipes, creep tests are performed and are extrapolated to much longer periods of

Table 1.5 Phenomena responsible for potential change of strength and stiffness in plastics.

Phenomenon	Maximum stress	Maximum strain	Stiffness
High-temperature thermal environment	+	+	+
Creep			+
Creep rupture	+	+	
Stress relaxation			+
Strain corrosion	+	+	
Moisture environment	+	+	+
Hygro-thermal effects	+	+	+
ESC	+	+	
Oxidation	+	+	
Loading rate	+	+	+
Ageing	+	+	+

pipe service life. For long-term extrapolation of the calculated creep rupture stress, regression analysis is used. Fig. 1.15 shows an example of long-term internal hydrostatic creep rupture behavior of PE-HD pipe at 40° according to software ADAP (automated design and analysis of pipelines) and the standard extrapolation method (SEM) based on the international standard ISO 9080.

The creep rupture curves related to the long-term internal hydrostatic pressure loading of pipes at different temperatures and media provide a very useful tool for the failure investigation and determination of the remaining service life of a piping system. For example, for the polyolefin-based thermoplastic pipes at elevated temperature three regions of long-term behavior may be identified (see Fig. 1.15). The first region characterizes the ductile response of the pipe and the third region signifies the brittle behavior caused by thermal and chemical ageing. The region in between these two is a transition region in which the qualitative changes in the pipe behavior occur. One of the standard tests in pipe failure investigations is testing of failed pipe samples at relevant temperature and medium and the comparison of the results with reference values. Depending on the position of the data points in one of the above-mentioned zones, a partial judgment about the condition of the pipe and the remaining service life may be made.

1.9.2   Ageing factors in pipe systems

Almost all activities in the area of life cycle engineering require knowledge of the time dependence and the ageing process of the system. For estimation of service life, the ageing mechanisms, the potential failures, the time to failure, and the reliability of the material and the system is to be estimated. *Ageing* of polymeric and composite materials

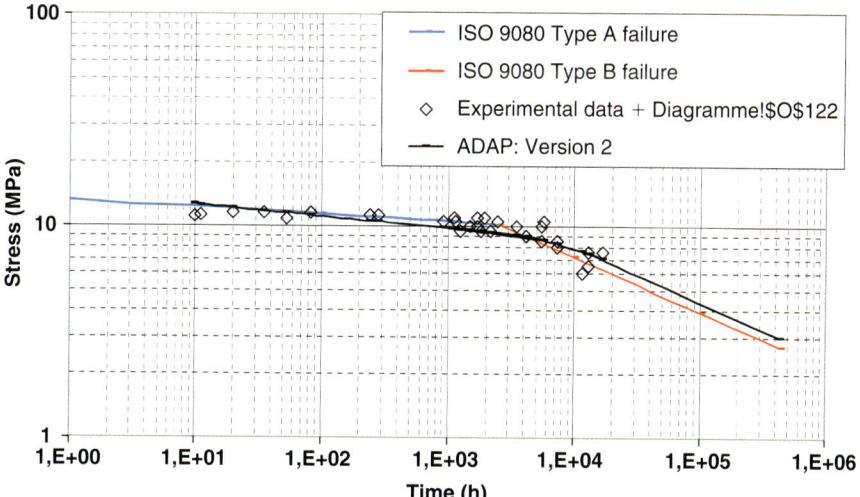

Fig. 1.15. Long-term internal hydrostatic creep rupture behavior of PE-HD pipe at 40° according to software ADAP and extrapolation standard ISO 9080: 2003.

is defined as the irreversible physical or chemical changes in the material and the material system. The following factors may contribute to the ageing of polymers and composites:

(1) Temperature effects (during production, processing, storage, transport, installation, and service).
(2) Time factor and its nonlinear interaction with temperature.
(3) Weathering: influence of the surrounding medium and, in particular, chemical agents (e.g., acids, alkalis).
(4) Stress: influence of mechanical actions, forces, loads.
(5) Biological factors.
(6) Other environmental effects (e.g., ultraviolet radiation).
(7) Interaction between mechanical, thermal, and chemical effects (e.g., ESC).
(8) Other material, system, and environmental effects such as incompatibilities, effects arising from the mode of application, and interference. Table 1.6 summarizes the factors influencing the ageing of polymeric materials and systems.

Table 1.6 Factors influencing the ageing and deterioration of plastics components.

Cause of ageing and deterioration	Factors
Weathering	(1) Radiation (solar, thermal)
	(2) Temperature (air, surface, extremes, alternating)
	(3) Moisture: fixed (frost, taw), fluid (rain, condensation, motion), vapor (relative moisture, wet/dry cycles)
	(4) Elements in the air (oxygen, ozone, carbon dioxide, gases like SO_2, NO_x, etc.), particles, fog, aerosol
	(5) See atmosphere
	(6) Underground conditions: ground water and contaminations
	(7) See water
Stress	(1) Static loadings: short and long term
	(2) Dynamic loadings: short and long term
	(3) Thermal conditions: short and long term
	(4) Residual stresses
Biological	Microbes, plants, insects, biological waste
Incompatibility	Mechanical, physical, chemical, thermal
Application	(1) Design
	(2) Production
	(3) Storage
	(4) Transport
	(5) Installation
	(6) Loading
	(7) Wear
	(8) Degradation

1.10 Reliability of pipelines

1.10.1 Concepts and definitions

1.10.1.1 Limit states

The service life and the failure events in material systems are quantified by the threshold values known as the *limit states*. A *limit state* is a state at which the structure can no longer satisfy the specified requirement. The limit states of plastic pipes are:

(1) *The ultimate limit state (ULS)*: This is a condition at which the limit of the strength of the pipe is reached. This state may be manifested by loss of water tightness, burst, and loss of stiffness.
(2) *The serviceability limit state (SLS)*: This is a condition at which the specified functional requirement of the pipe is no longer fulfilled. The manifestations of this limit are large deformations, change of color, buckling, clogging, abrasion, and local damages.

The condition of a pipeline as function of time can be qualitatively described by a monotonically decreasing curve, the so-called *P-F curve* (Fig. 1.16). On the P-F curve, a so-called *P-F interval* can be defined. The P-F interval is the interval at which the potential failure of the pipe will be realized as the actual failure in the form of the ULS or SLS.

1.10.1.2 Failure frequency

The so-called failure frequency is an index for the failure events in a piping system. It is defined as the number of kilometers per year or sometimes 1000 km/year.

1.10.1.3 Estimation of the failure risk

The failure risk and the expected risk can be defined as follows:

$$\text{Risk} = \text{Probability of failure} \times \text{Consequences}$$

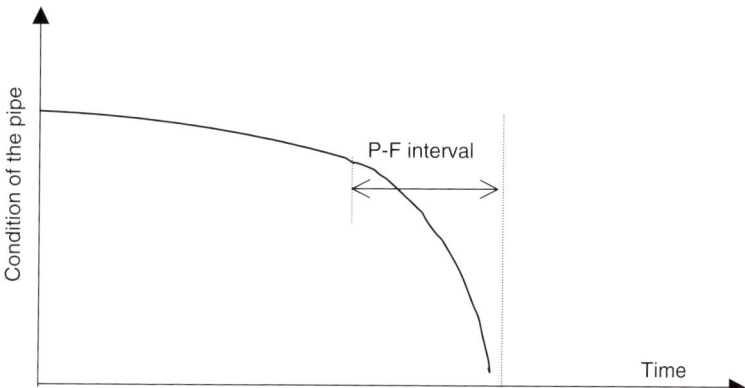

Fig. 1.16. The schematics of a P-F curve.

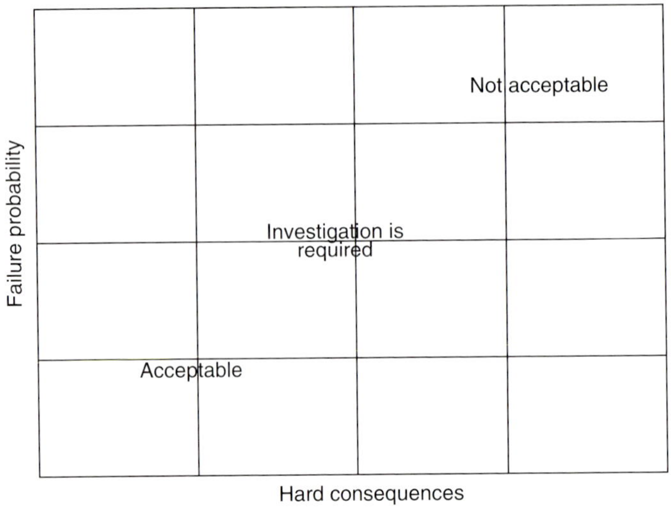

Fig. 1.17. Typical risk evaluation matrix.

Table 1.7 Stages of behavior of pipeline.

Stage	Measures
Behavior of pipeline	Failure-free behavior of pipeline at extreme conditions
Elimination of defects	Elimination of the elements or mechanisms that could cause failure
Evaluation of risk	Application of predictive methods
Preventive measures	Measure to prevent the failure before it occurs
Reactive measures in the case of failure	Measures to counteract when a failure has already happened

On the basis of this definition, a failure-risk matrix can be constructed (Fig. 1.17). The appropriate measures can also be defined on this basis using the failure-risk matrix.

Table 1.7 summarizes the stages of treatment of failure cases including the risks and the required measures.

1.10.2   Cost-use analysis

The risk matrix can be used as a basis for the qualitative cost-use analysis. In the matrix shown in Fig. 1.18, the level of expected costs is qualitatively related to the degree of the involved risk.

1.10.3   Risk management system for pipelines

The risk management of a piping system is a system of activities that contains technical, financial, environmental, and social aspects as its subsystems. The elements of a risk management system for pipelines can be identified as follows:

(1) Overall risk management system
(2) Objectives

Fig. 1.18. Cost-use analysis matrix.

(3) Description of the pipeline and the legal aspects
(4) Organizational aspects
(5) Key personal, qualifications, responsibilities
(6) Interested bodies and circles
(7) Documentation and communication system
(8) Management of change
(9) Life cycle analysis
(10) Integrity management (controls, repairs, . . .)
(11) Emergency planning
(12) Emergency measures
(13) Execution measures
(14) Management system review
(15) Audits, feedbacks, implementation

1.11 General procedure for reliability assessment

The main goal of health diagnosis and failure assessment is to find the cause(s) of failure, the sources of failure, and the reliability of the piping system. The reliability assessment can lead to a number of corrective actions related to the piping system. Following a failure investigation or a reliability assessment of a piping system, several strategies may be adopted on which the further actions can be planed. From the technical point of view, when confronted with a pipe failure, one of the following attitudes may be adopted:

(1) Zero scenario (no action).
(2) No action, but monitoring the situation.

(3) An ad hoc preliminary ambulant investigation and actions including repairs and replacements.
(4) Comprehensive failure and reliability assessment and rehabilitation.

A comprehensive health diagnosis and failure and reliability assessment of pipelines require a systematic procedure. It involves the following stages:

(1) Legal clarification of the failure case and the degree of expert involvement.
(2) General orientation about the failure case.
(3) Setting up of questionnaire, which should be put together by the involved parties? This questionnaire contains the extent of activities and details of the issues to be assessed.
(4) Visual observation of the object, site visits, and interviews.
(5) Documentation of the failure event.
(6) Collection, organization of available information.
(7) Study of the history of the case, the repairs, changes, . . .
(8) Assessment of environmental effects, requirements, function.
(9) Building up rudimentary hypotheses and preliminary assessment.
(10) Planning of necessary investigation, including laboratory tests and field investigations.
(11) Calculations and modeling.
(12) Synthesis of results.
(13) Validation of hypotheses.
(14) Choice of the most plausible hypothesis.
(15) Reliability and risk analysis.
(16) Responding to the questionnaire regarding the failure causes and failure sources.
(17) Suggestions and recommendations.
(18) Reporting of the result of investigation.

1.12   General safety questions related to pipe systems

As mentioned before, pipelines in general are one of the most reliable means of transport of materials and energy. Among various types of piping systems, plastic pipes enjoy the advantages of lightweight, corrosion resistance, hydraulic efficiency, relative ease of flexible connections, and installation. Statistically speaking, all world systems may experience occasions of malfunction and failure. In recent years, plastic pipe systems have proved to be statistically quite safe with relatively low record failure incidents. Nevertheless, for proper functioning of these modes of transport, a general discussion of safety of pipelines including plastic pipes is quite relevant.

From the technical point of view, the question related to the safety of piping systems is related to the fulfillment of functional requirements on the system. In this sense, any malfunction can be regarded as the reduction of safety. In this connection, the following technical questions may be raised:

(1) What categories of dangers could be caused by the lack of pipeline safety?
(2) What factors or mechanisms have caused the immediate and potential lack of safety?

(3) What is the extent and importance of the lack of safety and what are the consequences?

(4) Which actions and measures should be taken to prevent the danger?

(5) Which corrective measures should be implemented to habilitate the piping system and to prevent future potential failures?

(6) What are the safety limits of the existing piping system?

A systematic consideration of these questions should include quantification of the safety parameters and factors. Various means of quantification are on-site measurements, laboratory testing of samples, engineering calculations, and periodic instrumented inspections. The existing pipelines might have been planed for the requirements and loadings, which might have changed. Moreover, the plastic pipe material had also experienced some degree of ageing and damages. Furthermore, the standards governing the pipes and the related safety factors might have changed. All these probabilities necessitate a proper reexamination of the safety questions and reassessment of reliability of the existing piping system. The failure event, which may occur in the pipeline, should be a signal on the partial malfunction of the system. Hence, a proper examination of the event would provide a very useful means of the assessment of the safety of the pipeline and a motivation for the health monitoring and assessment of the remaining service life of the system. An objective and systematic failure investigation is a constructive action that may prevent much severe damages.

2
Failure investigation of plastic pipes

2.1 Failure phenomenon

Statistically, pipelines generally have a better safety record (deaths, injuries, fires/explosions) than other modes of transportation. In spite of this, piping systems like all other systems may experience certain events known as the failure event. The *failure event*, in a piping system is, by definition, a situation, which can hinder its function, change its configuration, jeopardize its integrity, and potentially endanger the environment. Failure mode is an event or mechanism that causes the pipe to reach one or combined strength and serviceability limit states defined in Chapter 1. Pipelines are considered one of the safest of the technical installations. A limit state function is a relation among the parameters that characterize a particular failure mode. Potential consequences of pipeline failures include loss of property, injuries and even fatalities, as well as costly property and environmental damage.

2.2 Sources of potential failures

A piping system may fail due to a variety of causes and damage mechanisms. The failure causes may include design errors, material failure, production, installation, service conditions, and environmental factors including mechanical loads, thermal effects, chemical agents, and ageing factors. Depending on the failure mechanisms a variety of modes of failure may occur, which include rupture or fracture, loss of stiffness, large deformation, instability and buckling, deterioration of properties, and functional inabilities.

2.2.1 Material deficiencies

Proper material selection is of prime importance for safety, serviceability, and durability of pipes. Polymeric materials have many positive features such as light weight, durability, processing feasibility, corrosion resistance, high stiffness to weight ratio, and ease of connection. However, there are some material aspects, which if not taken into consideration, may cause potential features. Time and temperature dependence of material properties, sensitivity to ultraviolet (UV) radiation and chemical agents, potential environmental stress cracking, and ageing are some of major salient features of polymers, which must be considered. In addition to these material properties, the emerging properties in the finished product such as residual stresses are of importance and should also be taken into account.

2.2.2 Insufficient design

Proper material selection alone is not enough to prevent the potential failure. Plastic pipes should be properly designed against all factors that may endanger the useful life of the system. For proper design of pipes, valid standards, guidelines, and computational methodologies must be used. One should always keep in mind that pipeline is not only an assemblage of materials or pipe sections, but it is a *system*; hence, in designing the pipelines, the systems behavior must also be taken into consideration. For example, it is important to consider all possible load cases in a buried pipeline including the longitudinal effects such as partial settlements, thermal changes, and directional changes. Hence, from this point of view, the available design methodologies according to which the dimensions of a local section of the pipe are calculated on the internal hydrostatic behavior only may not be considered as sufficient. Insufficient designs may lead to various modes of potential failure.

2.2.3 Problems related to processing

Many important properties of plastic pipes are dependent on the processing of the material and manufacturing of the pipe. For example, presence of impurities and insufficient thermal processing may lead to inherent material weakness. Moreover, residual stresses are features that arise from the processing of pipes and pipe connections.

In the case of fiber-reinforced plastic pipes, proper processing finds additional aspects; these may include the appropriateness of resin, type and properties of fibers, and certainly the processing of the material and product. Insufficient attention to any of these aspects may lead to an inherent material and product weakness, and may contribute to failure of the pipe.

2.2.4 Inappropriate storage, transport, and installation

A number of failures of plastic pipes can be related to an early life stage of the pipe. Improper storage may cause premature material ageing due to UV effects as well as storage damages and pre-deformations in the pipe itself. Also, during transport phase, damages and deformations may occur in the pipe. The role of proper installation of the piping system is also of outmost importance. Improper handling of pipe during installation, incorrect connection, improper bedding condition, and unpredicted loading conditions may all be factors, which may lead to an early life failure of pipelines. In the case of buried pipes, certain construction errors like compaction impact damage, uneven bedding, poor back filling material, poor connections, and improper supports may lead to initial weaknesses.

2.2.5 Unfavorable service conditions

There are numerous examples of pipe failure, which are attributed to the misuse of the material and the pipe, use of product beyond its intended lifetime, insufficient controls, and improper repairs. Moreover, unpredicted mechanical, thermal, chemical, electrical, and biological factors may cause potential failure. A number of failures can be related to the pipeline itself, while there are also failure cases, which may be caused by the interaction of the piping system with its environment. Maintenance damage can be considered as one of the causes of failure of pipes in service.

2.2.6   Third-party damage

One of the major threats to the safety of pipelines, particularly in and around the cities and the communities arises from individuals or organizations, referred to as *third parties*, carrying out activities in the vicinity of buried pipelines without realizing that the pipeline is there. The excavating equipment can accidentally strike and damage the pipeline. Severe damage can cause the pipeline to leak or rupture, resulting in an immediate hazard to property and life. Even the damages which appear to be minor can, over a period of time, weaken a pipeline and cause leakage or rupture in a later period. Damage can also result from large or small excavation projects, caused by a contractor during, for example, constructing a building or a road, adjacent construction, drilling, and impact loading. The intended damage and sabotage actions are the intervention in the piping installations, specially exposed pipelines, which should by no means be underestimated.

In an objective failure investigation, the possibility of the third-party damage should be kept in mind, but should not be overemphasized. One should note that the "outside damage" is not the same as "third-party damage." The outside damage is simply a damage done to the *outside* of a pipeline; the company's employees or contractors often cause such damage. The employees or contractors of other linear facilities (such as railroads or natural gas pipelines) sharing a right-of-way with a pipeline may also cause such damages. One should be conscious that it is always easy to blame others for a failure event; to substantiate or to refute such claims, however, an objective failure investigation should be carried out.

2.2.7   Ageing and deterioration

Polymeric materials age and their properties deteriorate in the course of time. Mechanical, thermal, chemical, biological, and UV rays effects can accelerate the process of deterioration. The fact that thousands of kilometers of pipelines have been installed since several decades and the fact that several generation of plastic materials has been used in the existing pipelines, which may not necessarily meet the new requirements provides a reason for considering the ageing as one of the sources of potential failure of the existing pipelines.

2.3   Types of failures

Failure of pipelines is, in nature, a combinatory statistical phenomenon, that is an event which may have several causes in a probabilistic medium. The failure event itself may be manifested in several modes known as the *types of failure*. Fig. 2.1 shows an overview of various failure modes and related failure mechanisms. In a general classification, failure types may be divided into following categories.

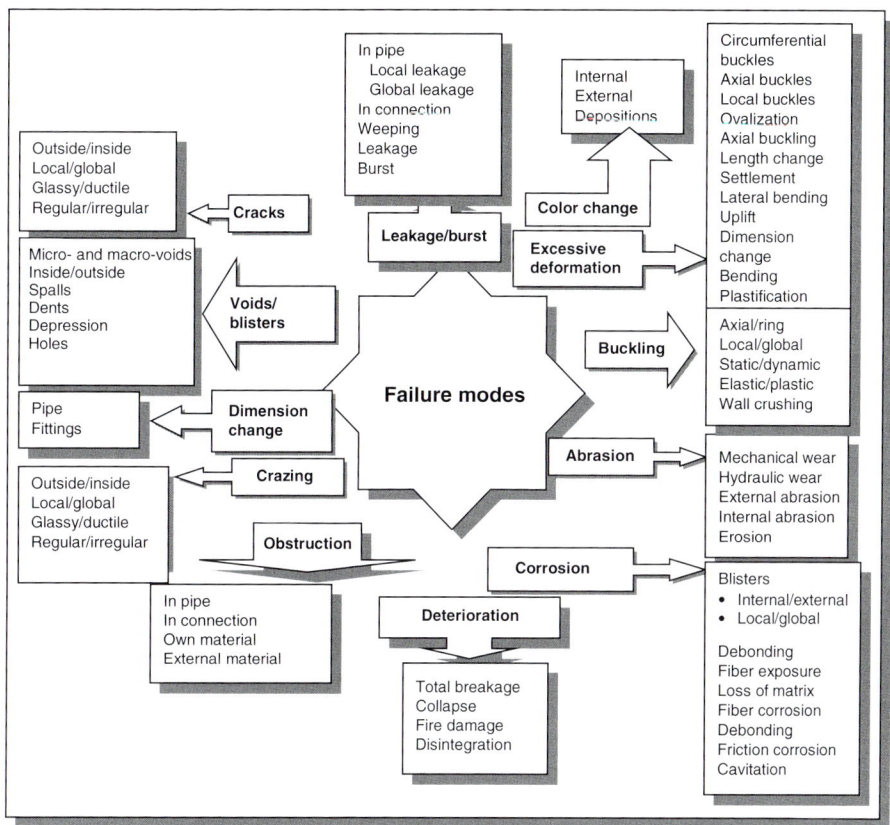

Fig. 2.1. Potential failure modes of piping systems.

2.3.1 Mechanical failure

Mechanical failure arises from applied external forces that exceed the strength or the maximum strain capability of the material in the sense of the strength limit states. The mechanical failure may be manifested by cracking, burst, breakage, and other types of loss of load carrying capacity. It may occur in short term or in longer time periods.

2.3.2 Thermal failure

Thermal failures occur from exposing the pipes to an elevated temperatures or extremely cold environment. Plastics tend to get brittle at low temperatures. At very high temperatures the pipe may warp, twist, melt, or even burn. Thermal ageing occurs at the pipes at elevated temperatures. One of the manifestations of thermal ageing is embitterment of material and occurrence of cracks.

2.3.3 Chemical failure

Chemical agents may affect the material properties. One of the phenomenon is known as chemical ageing in which the material, physical, and surface properties undergo deterioration. Residual stress, high temperatures, and external loading tend to aggravate the problem of chemical ageing.

2.3.4   Environmental failure

Polymeric material and products exposed to aggressive environments are susceptible to various types of detrimental phenomena. UV radiation, humidity, microorganisms, ozone, heat, solvents, oils, and pollutions are major environmental factors that seriously affect the properties of polymers. The effect can be manifested by the change of color, crazing, and cracking. These effects may even lead to a complete breakdown of the polymer structure.

2.3.5   Brittle and ductile failures

Plastic materials such as polyolefin-based thermoplastics and the pipes made of these materials have, in general, quite a high degree of ductile behavior at room temperature and low speed of loading. However, even highly ductile material such as polyethylene may demonstrate a pronounced brittle behavior. For example, low-temperature and high loading speed together with residual stresses may lead to a brittle response of these otherwise ductile materials and the products. The phenomenon of rapid crack propagation (RCP) is one of the manifestations of the brittle fracture failure.

2.4   Failure mechanisms in polymeric products

Failure of polymer materials and products may be viewed at various scales ranging from molecular to macroscopic viewpoints. Table 2.1 summarizes the environmental agents responsible for failure in polymeric materials and plastic products in the molecular level. It also gives an overview of the resulting mode of failure and the general appearance of the material. The agents listed in this table include mechanical, thermal, chemical, biological, and time factors.

2.5   Failure analysis

Failure analysis is the science and technique of understanding how materials and products fail. Whenever a component no longer performs its intended function, it is valuable to understand how and why it has failed. Failure analysis is a critical part of understanding what went wrong, what could have been done to prevent the failure, and how one might prevent similar failures. Any type of failure is, in nature, undesirable, but in some applications the consequences may go beyond inconvenience or cost, and may become critical or life threatening to the extents that even one failure, no matter how rare, would be considered unacceptable. Past experience shows that most failures are not extraordinary events; they are often foreseeable and some are quite preventable.

One should not always look at the negative side of failure analysis as an activity, which tries to find the wrong things in material, design, production, installation, and service. On the contrary, an objective failure analysis can lead to several positive consequences:

(1)   A valuable knowledge base, which can be used to great advantage in preventing further failure events
(2)   To improve the products and services

Table 2.1 Effects of various agents on failure of polymeric materials.

Effect	Mechanism	Result of the action	Appearance
Mechanical forces	Mechanical weakening of molecular valances	• For amorphous polymers (e.g., PVC) no special optical indications • For part crystalline polymers, (e.g., PE) stretching of fibrils	• Relatively flat surface with little or no traces of deformation • Non-flat surface, traces of deformation
Thermal effects	For part crystalline thermoplastic materials, at high temperature, softening of physical bonds and then melting of non-cross-linked thermoplastic polymers	Plastic deformation	Formation of faults voids, and other melting structures; for example, columnar forms
Chemical effects	Weakening of molecular valances by solvents. Penetration of the solvent into the polymer resulting the swelling of polymer.	Formation of crazes (micro-cracks due to normal stresses, stretched fibrils having diameter of 0.01 to 0.1 μm) on the surface or inside the object.	Formation of shrinkage faults due to diminishing of swelling through vacuuming of the material or by the evaporation of solvent. The crazes usually have a round formation emanating from a center.
Ageing	Thermal degradation, Stress corrosion	Embitterment, change of color	

PE: polyethylene; PVC: polyvinyl chloride.

(3) To increase the useful life of pipeline installations
(4) To optimize the economic factors
(5) To increase the safety of pipeline installations

Failure analysis of pipe systems always requires careful observation of many factors including the planning, the material, the installation, the service conditions, the repairs and intervention, and the environmental effects.

2.6   Why failure analysis?

The failure investigation of pipelines and determination of failure causes may be an objective by itself, which could be of interest for the involved parties. It serves, however, to fulfill other goals including rehabilitation, failure prevention, and technical improvements in the material and the system. Hence any failure investigation should not be looked upon as an activity that tries to devaluate certain material classes, products or producers. An objective failure investigation can even support the parties involved in the failure event in the sense of providing them with a new knowledge that can be used to improve their products and services. From this point of view, the major goals of a comprehensive failure analysis are:

(1)   Determination of the failure cause (diagnosis)
(2)   Analysis of the state of a pipeline (heath monitoring)
(3)   Statement about the future trend of the pipeline (prognosis)
(4)   Estimation of the remaining life (reliability analysis)
(5)   Guide for rehabilitation (road-map)
(6)   Maintenance on the basis of the health monitoring (retrofitting)
(7)   Prevention of similar failures (learn effect)
(8)   Correct planning of new pipelines
(9)   Increase of safety of the existing and planed pipeline systems

2.7   Parallel between engineering and medical diagnosis

There is indeed a parallel between the two branches the *medicine diagnosis* and the *engineering diagnosis*. Both of these two disciplines utilize related scientific basis together with the data and the devices as well as the knowledge gained in practice to arrive at the plausible causes of malfunction of the system. Obviously, the biological systems are more complicated than the physical installations such as pipelines. The methodology of prognosis is however the same in both disciplines and the parallel viewpoint quite useful. Fig. 2.2 show the parallelism between the engineering and the medical disciplines.

2.8   Terminologies related to failure investigation

The terminologies related to failure analysis used by various experts in different countries are not exactly the same. However, for coherent and efficient dialog among the experts, a common terminology may prove to be helpful. The following is a list of some terms related to failure investigation of materials and structures.

- *Failure*: Changes in the system that reduces or destroys the conceived function of the system or provide the grounds for such situation to occur.
- *Previous failure*: Failure that previously occurred in an element or in the system.
- *Primary failure*: The failure, which occurs first and the cause of further failures.
- *Consequential failure*: Failures, which occur due to a previous failure in the same element or another element of the system.
- *Recurrent failure*: Recurrence of failures of the same type.
- *Failed element*: The element or a piece of an element affected by a failure.

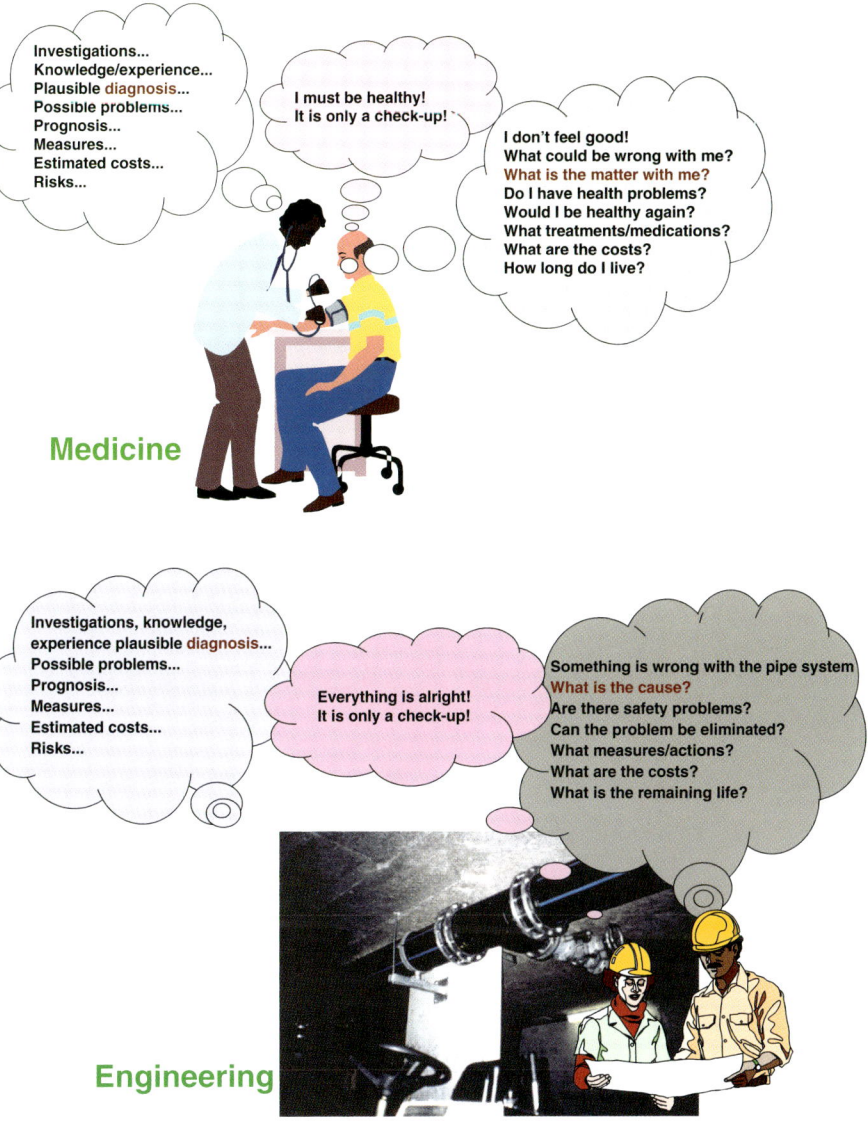

Fig. 2.2. A comparison between medicine and engineering diagnosis.

- *Failure zone*: The region in the system in which the failure has occurred.
- *Failure picture*: The appearance of the failed element.
- *Failure features*: Main characteristics of the failure phenomenon.
- *Failure process*: Temporal development of a failure phenomenon.
- *Failure type*: Arrangement of a failure event according to the failure categories.
- *Failure cause*: The influence or influences triggering the failure.

- *Failure hypothesis*: The probable cause(s) and course of events in a failure phenomenon.
- *Failure analysis*: Systematic investigation of a failure event to determine the cause(s) of failure.
- *Failure correction*: Measures to prevent the occurrence of the same type of failure.
- *Failure prevention*: Preventive measures against the occurrence of failure.

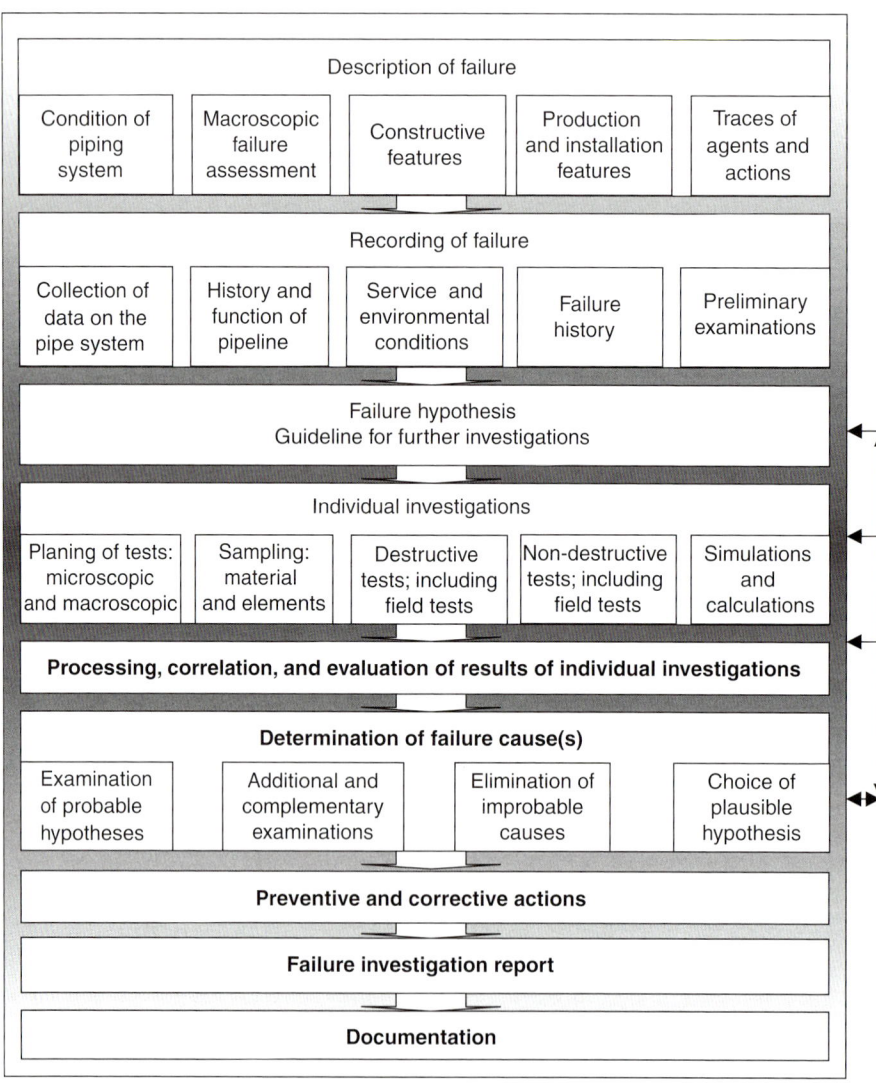

Fig. 2.3. Steps in a failure investigation.

2.9 Steps in failure investigation

In any failure and reliability assessment, the first rational step is to make a systematic documentation of the system, its history, and the failure event as well as the immediate consequences. In fact, the assessment of failure starts at the stage in which a failure is observed and properly documented. The documentation of a failure event includes, protocols of the site visits, the meetings, interviews, gathering of records, documentation of the object and its surroundings, sketches, the fist analyses and assessments, and the measures taken or the intended actions. Further steps include planning of required investigations, examination of hypotheses, and determination of plausible failure causes, measures to reduce the potential damages, corrective actions, and appropriate documentations. Fig. 2.3 shows the necessary steps in a failure investigation. Depending on the type of investigation, some of these steps may be eliminated or enhanced to include additional activities. The legal and economic aspects are not included in these steps.

An important aspect in a failure analysis is the role of *feedbacks*. A failure investigation is not a "one-way" street starting from failure description and ending in a failure judgment. The process of a failure investigation should have a *learning component*. This feature can be embedded into the failure investigation methodology by implementation of several feedbacks. At any stage of failure analysis the need may arise that one should *go back* to the earlier stages and make modifications in procedures and conclusions. Fig. 2.3 shows this feature by arrows that go back and forth among several steps of failure analysis. It is also very important to keep in mind that in carrying out all of the investigations, safety measures must be taken. These include using proper equipment and safety devices, wearing appropriate clothing and taking precautions for undesired or unexpected events during investigations. Details of the steps suggested in Fig. 2.3 are elucidated in Table 2.2.

2.10 Details of steps in failure investigation

The stages of failure investigation outlined in Fig. 2.3 include more detailed activities, which should be followed in a parallel or a sequential fashion. Table 2.3 outlines these activities at each step without a preferred sequence. The priority of each action should be decided upon in each particular case. Furthermore, depending on the failure investigation case, additional activities may also be needed.

One of the major steps in a failure investigation is *documentation* of failure event. Documentation of a failure event includes, protocols of site visits, meetings, interviews, witness reports, gathering of records, documentation of the object and its surroundings, sketches, the fist analyses and assessments, and the measures taken or the intended actions. For this purpose, a systematic procedure is required. Table 2.4 is an outline of a proposed protocol for documentation of failure event. A failure investigator can use this table for a systematic documentation.

Based on the first series of documentations and data gathering, a number of hypotheses about the failure cause(s) may be created. Some of these hypotheses may be refuted in later stages; the rest may be kept and examined during further steps of investigation. To build up plausible hypotheses a set of intelligent questions and examinations about

Table 2.2 Details of steps in a failure investigation (outlined in Fig. 2.3).

Category of investigation	Steps in investigations	Means of investigation	Examples
Description of failure	• Identification of the case	Site visits, interviews, reports of failure	Functional, non-functional, global or local failure
	• Number and frequency of failures		
	• Onset of failure		
	• Failure event (sudden, gradual)		
	• Events and interventions prior to failure		
	Constructive features	Site visit, drawings, plans	Connections, supports
	Macroscopic failure identification	Site visit	Deformations, cracks
	Production and installation feature	Site visit, drawings, plans	Material type, bedding condition
	Traces of other agents	Site visit	Oils, scratches
Recording of failure event	Information about the pipe system	Site visit, drawings, plans, specifications, test reports	Dimensions, materials, connections
	History of the pipeline system	Archives, journals, repair reports	Date of installation, number of repairs
	Function of the pipeline	Documents, questionnaires	Gas, water, drainage
	Service conditions	Site visits, interviews, field measurements	Internal pressure, traffic, water hammer, chemicals
	Failure process	Reports of failure	Time, mode of failure
Failure hypothesis	Use of failure guidelines	Books, literature, standards, reports	
	Examination of proposed hypothesis	Interviews, study of failure protocols	
	First expertise	Use of previous experience, use of failure data on similar piping systems	
	Classification of hypothesis		
	First evaluation of hypothesis		
Individual investigations	Planning of investigations	Site sampling, labor sampling	Material tests, field tests
	Sampling:		Tensile test samples
	• Planning for sampling (time, position, size, number, distribution)		
	• Sample identification (marks, designations, . . .)		

• Documentation of extracted samples • Conditioning of samples • Testing		
Destructive tests • Planning of the tests • Choice of proper apparatus • Choice of appropriate testing standard • Determination of test parameters • Execution of tests • Interpretation of data • Estimation of error	Mechanical, physical, and chemical tests; testing of reference materials and objects, if possible	Burst tests, long-term tests
Non-destructive tests Simulations and calculations Determination of material parameters and comparison with the reference data	Ultrasonic, acoustic emission Analytical modeling, numerical simulations (e.g., finite element analyses), software (ADAP, finite element, etc.) standards	Internal pressure loading Stress, strain and deformation analysis Buckling investigation Simulation of the system behavior
Processing, correlation, and evaluation of individual investigations	Results of site investigations, laboratory tests, simulations and calculations	Strength limit of materia.
Determination of the failure mechanism and failure cause	Set up possible failure hypotheses Examination of probable hypotheses Repetition of investigations/complementary investigations Verification and elimination of improbable causes Choice of the most plausible hypothesis	Synthesis of results	Material failure, service problems
Measures to be taken	Expertise, experience, possibilities, extent of damage	Local repair, total replacement
Failure investigation report		

ADAP: automated design and analysis of pipelines.

Table 2.3 Documentation of failure event.

Category	First hypotheses and actions	Notes
Failure mode(s)	(1) Crazing (2) Cracks (3) Excessive deformation (4) Buckling (5) Leakage/burst (6) Color change (7) Dimension change (8) Blisters/voids (9) Corrosion (10) Abrasion/erosion (11) Obstruction (12) Deterioration/disintegration (13) Other	
Hypotheses about failure cause(s)	(1) Mechanical (2) Thermal (3) Chemical (4) Biological (5) Other (state)	
Hypotheses about failure source(s)	(1) False planning (2) Insufficient dimensioning (3) Defect in material (4) Error in production (5) Storage errors (6) Transport errors (7) Installation errors (8) False application (9) Operational errors (10) Errors in repairs (11) External interference (12) Other (state)	
Actions undertaken	(1) Nothing at all (2) Nothing for the time being (3) Testing (4) Expertise (5) Assessment (6) Preliminary repairs (7) Permanent repairs (8) Relining (9) Renewal of the pipeline	
Further actions	(1) Nothing (2) Detailed investigation (failure and reliability assessment) (3) Actions without further technical investigation (4) Non-technical measures (especially insurance) (5) Other	
Persons involved in assessment	(1) Technically untrained (2) Technically trained (3) Experts	

Table 2.4 Guiding questions related to the failure and reliability assessment of pipes.

Environment	Guiding questions
Mechanical	Mechanical forces?
Internal pressure	Pressure pipe?, overpressure?, water hammer?
External pressure	Buried Pipes?, Pipeline on see the floor?, high traffic load?
Axial tension	Hanging pipeline?, axial load bearing joints?, shrinkage?, temperature?
Axial compression	No length change compensation? fix point?, pipeline on slope?, temperature?
Bending	Any type of bending action?
Traffic load	Buried pipe?, change in the traffic load?, extra overburden?
Settlement	Soft bedding?, new installation?
Uplift	Ground water?, massive adjacent objects?
External interference	Repairs?, construction works?, intended damage?
Impact	Impact during transport?, construction works?, intended damage?
Vibration	Earthquake?, bridges?, vibrating machines?
Fatigue	Cyclic loading?, vibrating machinery around?, pipe on bridge?
Residual stresses	Uncontrolled cooling during production?
Other mechanical actions	
Thermal	Thermal effects?
High temperature	Operational temperature?, storage temperature?, solar radiation?, external heat sources?
UV radiation	Direct exposure to solar radiation?, UV lamps around?
Fire	External fire?, internal fire?
Frost	Low inside or outside temperatures?
Other thermal effects	
Chemical	Chemical effects?
Water	Water from outside?, water inside?, ionized water?
Oxygen	In the air?, in operation?, air content in the fluid
Acids	Outside?, inside?
Alkalis	Outside?, inside?
Solvents	Outside?, inside? what type of solvent?
Oils	Outside?, inside?
Benzene	Outside?, inside?
Other chemical agents	
Operation/service	Change of application?, lack of maintenance?
Abrasion	Fluid alone?, fluid containing solid particles?
External interference	Repairs?, construction works?, intended damage?
Other service conditions	
Biological	Biological effects?
Microbes	Sign of microorganisms?, mushroom?
Insects	Sign of insects?
Other biological factors	
Time factors	Long in operation?

the failure event may prove to be quite helpful. Table 2.4 is a guideline for the type of technical questions which may be raised by the failure investigator.

2.11   Testing of plastic pipes

Experimental investigations are an important element of a sound failure analysis. Planning of relevant experiments, appropriate performance of tests, their interpretations, correlations with theoretical simulations, and comparison with the observed phenomenon are essential steps, which play a major role in the right judgment about the failure mechanisms and causes.

Testing of plastic pipes include material tests, product tests, and tests on the pipe system. Some of these tests have destructive character in the sense that through testing the sample or the element will be broken or disintegrated. The other types of tests are non-destructive and after testing the original properties of the material or the system are preserved. Testing may occur in the field or in the laboratory atmosphere.

Categories of tests for plastic materials and pipes include:

(1) Macroscopic and microscopic examinations (at various magnification levels)
(2) Physical tests
(3) Mechanical characterization
(4) Chemical analysis
(5) Thermal tests
(6) Biological tests
(7) Fire tests
(8) Tests on the element and the system (e.g., buckling and internal pressure testing)

For the material and element tests guidelines and standards are available that give the detail of the procedure to be followed in each testing. In failure investigations, choice of appropriate standard and testing methodology is part of a correct planning that can even influence the result of the failure analysis. In the cases in which no standard is available, new tests should be planned; test devices should be built, calibrated, and used with sufficient degree of relevance and confidence.

It is not possible to list all of the available tests on plastic materials and pipes. For reference purposes, however, a short list of some important tests is useful. Table 2.5 presents a list of the tests and the results to be expected from the tests. This table can be used as a quick reference for failure investigations.

2.12   Failure modes and potential causes

Presentation of various failure modes, their features, and related failure mechanisms in a compact form is useful. However, it should be kept in mind that not all relevant aspects can be included in such compact presentation. Nevertheless, as a means of easy reference and guideline for the user a tabular representation of salient features of plastic pipe failures is presented. Table 2.6 shows a compact presentation of failure modes, their features, and related failure mechanisms in plastic pipes. Examples and case studies related to each class of failure mode shall be presented in future chapters.

Table 2.5 A list of tests performed on polymers, plastics, and plastic pipes (not exhaustive).

Type of the test	Tests performed	Properties determined
General physical properties	Specific gravity tests	Specific gravity
	Density tests	Density
	Water absorption tests	Water absorption
	Moisture analysis	Determination of moisture content
	Sieve analysis	Particle size
	DSC, TGA, and TMA tests	Thermal properties, melting point, degree of crystallinity and glass transition
	DSC	Difference of thermal consumption of the test material with an identical reference material
	DTA	Difference of thermal behavior of the test material with an identical reference material
	TG	Heating of the sample and determination of the change in its mass as function of time and temperature
	Residual stress tests	
	Gas diffusion tests	Resistance to gas diffusion
	Hardness tests	Determination of hardness
	Dilatometer	Determination of dimensional changes of the object as function of temperature
Material characterization tests	Melt index tests	Flow properties
Chemical properties	FTIR, chromatographic techniques such as GLC, HPLC, GC, etc. and other spectroscopic techniques such as MS, NMR and UV-VIS	Chemical composition of the base material, and any additives, fillers, and possible contaminants
	OIT	Determination of diffusive properties of pipe material
	Capillary rheometer examinations	Rheometrical properties
	Viscosity tests	Determination of viscosity
	Tests for contamination	Identification and quantification of residual solvents
	Flammability tests	Determination of the degree of fire resistance

(*Continued*)

Table 2.5 (*Continued*)

Type of the test	Tests performed	Properties determined
Surface properties	Light microscopes, transmission electron microscopes SEM, magnification up to approximately 20,000X	Microscopic examination of fracture surfaces
Mechanical properties	Hardness	Hardness properties
	Tensile test	Tensile properties (strength, modulus, maximum strain)
	Bending test	Bending properties (strength, modulus, maximum strain)
	Compression test	Compression properties (strength, modulus, maximum strain)
	Creep tests	Long-term properties, creep module, . . .
	Relaxation tests	Stress relaxation modulus
	Charpy test, Isod test	Impact properties
	Fatigue tests	Fatigue resistance, number of cycles to failure
	Slow crack growth tests	Resistance against slow crack growth at various temperatures and different medium
	Fracture mechanics tests	Determination of fracture toughness
	Abrasion and wear tests	Abrasion properties, wearing properties
Thermal tests	Thermo analysis	Coefficient of thermal expansion
		Determination of thermal, gravimetrical and dilational effects in material as the result of physical or chemical changes
		Thermal conductivity
		Determination of specific heat
Electrical tests		Dielectric strength
		Dielectric constant and dissipation factor
		Electrical resistance
		Arc resistance
Weathering	Accelerated weathering tests	Weathering properties
	Fungi tests	Resistance to fungi

Property	Test	Description
Optical properties	Bacterial tests	Resistance to bacteria
	UV tests	Resistance to UV rays
		Refractive index
		Luminous transmittance
		Photo-elastic properties
	Color tests	Color properties
Chemical properties	Immersion tests (acid, acetone, . . .)	Chemical resistance
		Stain resistance
	Solvent stress cracking test	Solvent stress cracking resistance
		Environmental stress cracking resistance
Product tests	Fractography (inspection of fracture surface)	The nature of the fracture (e.g., Brittle versus Ductile)
	Short-term burst tests	Determination of short-term burst resistance of pipe under internal hydrostatic pressure
	Long-term internal hydrostatic pressure tests	Long-term properties, creep rupture behavior
	Short-term and long-term pipe ring tests (pipe ring sections under diametrical forces)	Determination of the pipe ring stiffness in the air, in wet condition, and in other medium
	Strain corrosion tests	Determination of the pipe resistance to aggressive medium (such as acids environment)
	Slow crack growth tests	Determination of the fracture toughness of the pipe in various environments
	RCP tests	Determination of dynamic fracture toughness
	Buckling tests	Determination of buckling resistance of pipe under external hydrostatic pressure
	Non-destructive tests	Determination of structural integrity, determination of in-situ properties

DSC: differential scanning calorimetry; DTA: differential thermal analysis; FTIR: Fourier transform infra red; GC: gas chromatography; GLC: gas liquid chromatography; HPLC: high-performance liquid chromatography; MS: mass spectrometry; NMR: nuclear magnetic resonance; OIT: oxygen induction tests; SEM: scanning electron microscopy; TG: thermo-gravimetry; TGA: thermo-gravimetric analysis; TMA: thermo-mechanical analysis; UV–VIS: ultraviolet–visible spectroscopy.

Table 2.6 A compact table of various failure modes and their potential causes.

Type of failure	Sketch of the failure mode	Features of the failure mode	Potential causes
Crazing		White region, micro-cracks	Internal pressure, point load, chemicals
Longitudinal cracks		Through cracks, cracks on the inner or the outer side	Internal pressure
Ring cracks		Complete ringsor a sector of a ring	Bending, thermal effects
Irregular cracks		Irregularclusters of cracks at the inner or the outer side	Environmental stress cracking(stress + medium)
Mixed cracking		Through orsurface cracks, cracks on the inner or the outer side	Internal pressure bending
RCP		Through brittle cracks with or without branching	Dynamic loading at low temperatures
Wall section of a pipe with RCP		Glassy surface behind the crack front	

Mixed modes

Total fracture of the pipe	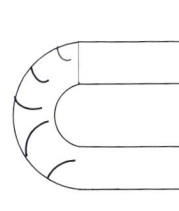	Internal pressure, external intervention
Ring cracks in a bent pip		Bending (low-bending radius)
Cracks in a concrete-embedded thermoplastic pipe	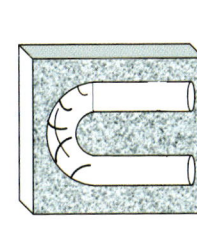 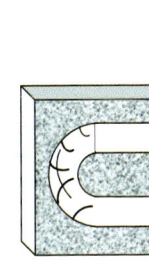	Bending, environmental stress cracking (stress + medium)
Buckling (ovalization)	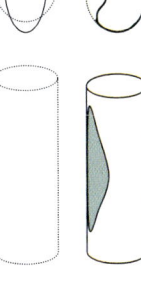 	External pressure, bending
Snap-through	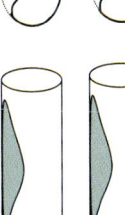  	External pressure, hard external bending
Two-sided snap-through	   	External pressure
Non-symmetric buckling	 	Longitudinal bending, Brazier effect

(Continued)

Table 2.6 (*Continued*)

Type of failure	Sketch of the failure mode	Features of the failure mode	Potential causes
Longitudinal buckling		Non-symmetric buckling shape	Axial compression, thermal effects (thermal buckling)
Buckling of the three layer pipe inside layer of a		Delamination of the inner difference in middle layer, layer from the usually accompanied by large inward buckling deformation	High temperature, coefficient of thermal expansion
Buckling of concrete-embedded pipe		Non-symmetric inward buckling of pipe	External hydrostatic pressure, settlement of pipe, heaving of the floor
Large longitudinal deformation			Partial settlement, heaving of the pipe
Change of color		Change of color near a butt fusion weld	Diffusion of gas through the weld
Delamination		Debonding of constituting layers	Weak bond, shear
Blazes		Relatively large packets of blazes; sometimes filled with fluid	Osmosis

Failure mode		Description	Potential causes
Voids		Localized or bundle of void spaces	Fabrication problems, entrapped air
Hole		Irregular though hole	External point load, impact, third-party damage
Abrasion and wear		Signs of erosion: sometimes exposure of fibers in GFRP pipes	Solid materials in the fluid medium
Obstruction		Debris inside the pipe	Foreign objects
Joint failure: butt fusion joint		Lack of tightness, separation of two pipe parts in the fusion weld	Weak fusion weld under axial, shear, and bending forces
Joint failure: electro-fusion joint		Lack of tightness, separation of two pipe parts in the electro-fusion weld	Axial tension, bending, internal pressure, improper welding
Joint failure: flange joint		Lack of tightness, separation of two pipe parts in the flange	Improper welding, axial tension, bending
Failure at the fixed point		Shearing of the pipe at the supports, global buckling of the pipe, local buckling	Shear action on the pipe

GFRP: glass-fiber-reinforced plastics.

Table 2.7 Features of various pipe connections and their potential weaknesses.

Technical requirements	Permanent (non-detachable) connections			Detachable connections	
	Weld connections		Glue connections	Cuff-link connection	Flange connection
	Butt fusion welding	Electro-fusion welding			
Long-term internal hydrostatic pressure strength	General reduction of strength and ductility at higher temperatures, internal pressures, and deficient welding	General reduction of strength and ductility at higher temperatures, internal pressures, and deficient welding	Long-term failure due to internal pressure and thermal ageing	Loosening of connection, loss of tightness of elastomer rings	Loosening of connection, loss of tightness
Strength of connection	Low long-term strength due to deficient welding under tension, torsion, and bending	Low long-term strength due to deficient welding under tension, torsion, and bending	Depending on the quality of glue and gluing	Potential failure due to axial forces	Depends on the quality of the connection
Tightness	Potential lack of tightness in presence of some fluid medium and gases	Potential lack of tightness in presence of some fluid medium and gases	Potential lack of tightness in presence of some fluid medium and gases	Potential lack of tightness in presence of some fluid medium and gases	Potential lack of tightness in presence of some fluid medium and gases
Vapor tightness	Potential lack of tightness in deficient welding	Potential lack of tightness in deficient welding	Potential lack of tightness in deficient gluing	Potential lack of tightness	Potential lack of tightness
Voids in connection	High risks in the case of deficient welding, potential of internal voids, surface notches and stress raisers	High risks in the case of deficient welding, potential of internal voids	High risks in the case of deficient gluing	Low risk	Low risk

2.13   Qualitative risk estimation of plastic pipe connections

It was mentioned before, that any pipeline is a *system* consisting of related pipe elements, their connections, reservoirs, pumping stations, and possibility of other installations. The pipeline as a system has certain systemic behavior, which is resulted from this emergent integral entity. Partial settlement of a buried pipeline, thermal behavior of the piping system, and soil pipe interaction are examples of the emergent systems behavior. Therefore, any reliability and failure assessment of a pipe system should take into account not only the element behavior, but also the system feature as a whole.

Plastic pipes enjoy a versatile group of connections. The plastic pipe connections may be classified into two main categories: the detachable and the non-detachable connections. The particular connection types for plastic pipes are:

(1)   Butt fusion weld
(2)   Electric fusion weld with cuffs
(3)   Glue connection
(4)   Cuffs
(5)   Flanges

Each of these types of connections offer certain advantages and at the same time potential for probable failures. The experience from failure investigations has shown that, like any other plastic product, a large number of failures occur not in the pipe itself, but in the pipe connections. Table 2.7 presents an overview of various pipe connections and their potential weaknesses.

2.14   Lessons learned from pipeline failures

Looking back at the underlying concept of the bathtub theory discussed in the previous chapter, the failure event(s) at any stage of the pipe life cycle can be considered as a probability. This statement has been verified through the experiences gained from failure investigations. Table 2.8 shows a tabular form of the potential causes of pipeline failures. In future chapters, case studies related to some of these events shall be presented.

Table 2.8 Potential failures and failure sources of pipe systems.

When? (life stage)	Why? (sources and causes)
Conception	Inappropriate, false
Planning	Incomplete, false, no planning
Material	Material weakness, inappropriate application
Production	Mistakes during production, missing enough QA
Storage	Exposures to damaging factors, initial damages
Transport	Initial damages (e.g., due to impact)
Installation	Initial deformations and other damages
Service	Load cases, insufficient safety factors
Intervention	Unexpected load cases
Ageing	End of the life phenomenon

QA: quality assurance.

The case studies in Chapters 3 to 8 will reveal certain features that are common in many failure events. Lessons can be learned from these cases and use them in material and system improvement and in prevention of other potential failures. To summarize, appropriate material selection, suitable design, quality production, appropriate storage, transport, and installation, improved maintenance, materials, inspection, operator error reduction, leak detection and prevention, and damage control are far more important for reducing pipeline accidents.

2.15   Failure case studies

In the following chapters of this book, a number of failure cases related to plastic pipes are presented. The presentations take place in tabular form and in a compact fashion. The cases discussed have actually occurred in practice and can provide a guideline for similar cases. Furthermore, the lessons learned from these cases can prove to be of help in material developments and improvements in planning, design, production, installation, and service conditions. Due to compactness of presentations, many details of investigations are left out. Each case study is organized in the following tabular form:

(1) Designation and title of the failure case
(2) Pipe material and dimensions
(3) Description of the system
(4) Failed part
(5) Observed phenomenon
(6) Failure description
(7) Environmental conditions
(8) Time to failure
(9) Tests performed
(10) Other investigations
(11) Failure cause(s)
(12) Suggested corrective actions
(13) Photo documentation

Some of the failure cases included in each of the book chapters may have had other causes and hence could have as well been placed in other chapters dealing with another failure mode(s). Therefore, reference to these cases has been given in the related chapters.

Each of these actual cases was originally supplemented by a comprehensive documentation containing the details of the case. However, for brevity of presentation, details are not included in the outline of these failure cases. Moreover, the information and the evidences leading to the hypotheses and the failure judgments are not elucidated.

It is to be emphasized that the failure cases outlined in this section are statistical events, which have been caused by specific circumstances. The fact that a certain material or pipe type did fail would by no means reflect the weakness or the malfunction related to that material. The circumstances leading to these events could have also caused damages in other pipe materials and products as well. One should bear in mind that the material and pipe are only two factors among the many that may cause a failure to occur. Hence the cases outlined in this section should be treated in an educational and supportive perspective and should be devoid from any value judgment and generalization about the specific materials and products.

3
Fracture of plastic pipes

3.1 Fracture

Fracture of pipes is one of the most statistically common failure events that may manifest it in various modes and give rise to various side effects. The importance of fracture is due to the sequential events that may occur; some of which may have serious consequences. Loss of strength and loss of tightness are the technical consequences of pipe fracture. However, certain side effects such as explosions in gas pipes and flooding caused in the case of water and drainage pipe systems may have far graver consequences.

Every material has a certain resistance to fracture. The ability of a material containing cracks to withstand an applied load is called *fracture toughness*. Some triggering agents can increase the potential of fracture. These factors are called *stress raisers*. Initial cracks or flaws may act as stress raisers and can intensify the stress and cause a stress concentration. The discipline concerned with the behavior of materials containing cracks or other small flaws is called *fracture mechanics*.

Fracture of polymers is a mode of failure, which is the result of weakening of the molecular bonds, separation of molecular chains, and finally separation of the fracture surfaces. Fracture is a universal phenomenon, which may occur in various material bodies under various circumstances. In pipe systems, fracture may not only occur in plastic pipes, but also in pipe made of other materials such as steel pipes, iron pipes, and composite pipes.

3.2 Fracture modes

In a material object, one may identify three pure modes of fracture. Fig. 3.1 schematically shows the three types of fracture modes. The basic fracture modes are:

(1) The cracking identified by movement of the two crack surfaces normal to each other. This type of cracking is known as the mode I fracture or the opening mode. This type of cracking occurs mainly due to stresses in the normal to the crack surface and the crack direction.

(2) The shear mode or the sliding mode, known as the mode II fracture, consists of sliding of the crack surfaces relative to each other. In this case the forces producing the crack are parallel to the crack direction.

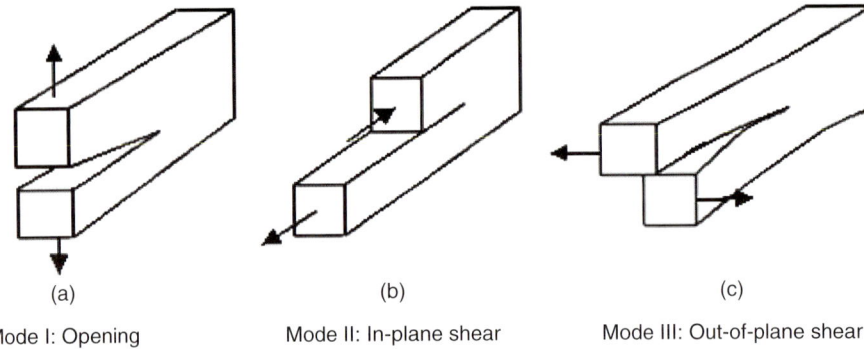

(a)   (b)   (c)

Mode I: Opening   Mode II: In-plane shear   Mode III: Out-of-plane shear

Fig. 3.1. (a) Mode I fracture: the forces are perpendicular to the crack; (b) mode II fracture: A in-plane shear crack; (c) mode III fracture: a tearing mode.

(3)   The tearing mode, known as the mode III fracture or an out-of-plane cracking, is a fracture which is identified by the lateral shear movement of the two crack surfaces relative to each other. In this case, the forces are perpendicular to the crack direction.

3.3   Fracture process

Fracture of plastic pipes may occur due to various factors; these include external applied loading, chemical agents, thermal gradients, production, and service conditions. In plastic pipes, fracture may occur in several stages; the main stages include:

(1)   Loosening of molecular bonds
(2)   Crazing
(3)   Formation of stationary cracks
(4)   Slow crack growth (SCG)
(5)   Crack propagation
(6)   Rapid crack propagation (RCP)
(7)   Total breakage

The fracture event is, in fact, a material stability phenomenon. An existing crack or notch can remain geometrically stable or may become unstable and grow. The initiation of growth is designated as fracture. For fracture to occur, the available energy should exceed the internal resistance of the material to create a new fracture surface. The equality of these two energies signifies the transition from a stable to an instable crack.

3.3.1   Crazing

Crazing is a name given to *fine surface cracks*, which may occur in plastics, especially amorphous polymers, under stresses smaller than the yield stress. Crazing usually occurs in the direction normal to tensile stress. The difference between a craze and a crack is that the crack is identified by separation of two surfaces while in the crazed region two surfaces are still bounded together by molecular fibrils (10 nm) which bridge two sides. Between these fibrils are micro-voids (10 nm). The cracked zone, on the other hand, does not contain a fibril structure.

The origin of crazing is believed to be a tri-axial tensile stresses at the molecular boundaries and discontinuities. High stress in impurities present in the material can produce macroscopic yielding. As the loading time increases, the ductile yielding occurs under smaller amounts of stresses. At low stresses, due to material defects, inclusions, local stress concentration regions, and residual stresses, microscopic yielding may occur. The micro-yield zones may coalescent and form micro-voids; these zones then fibrillate and become *crazes*.

A typical crazing zone can be divided into three parts: the fibrillar zone, the porous zone, and the rubbery zone. The rubbery zone lies at the very tip of the crazing. The tensile stresses are highest in the rubbery zone and are lowest in the porous zone.

Crazing can be accelerated through contact with fluidic- or gaseous-type aggressive medium including detergents. A tensile stress field especially in the presence of aggressive medium also accelerates the formation of crazes. If enough time is passed, more fluid is absorbed in the micro-voids. Due to fluid absorbency the yield strength of the material is locally reduced. As the result, a relatively small stress together with aggressive medium can cause crazing which may eventually lead to cracking.

3.3.2   Cracking

Cracking of pipes is a clear indicator of pipe failure. Cracking is actually a mode of fracture in the sense that the limit of fracture toughness is reached. Fracture toughness is a material property indicating the resistance of the material to the growth of imperfections, pores, and impurities. The definition as to what one may designate as "crack" is, however, quite wide and includes various scales of observation. In the molecular level, cracking may be considered as the weakening of atomic forces and partial or total disconnection of molecular chains. In larger scales, one may classify the cracks as *brittle* and *ductile*. In addition, some materials may demonstrate certain limited amount of ductility and may thus be designated as *micro-ductile* materials. One may refer to a behavior as micro-ductile if some degree of ductility in the microscopic scale is observed. Cracking may occur in one of the basic fracture modes discussed in Section 3.2 or it may occur in a combined mode.

3.3.3   Ductile versus brittle cracking

A classification of cracking into ductile and brittle can be useful for some failure investigations. From the macroscopic point of view, ductile cracking is the phenomenon in which the material undergoes plastic deformation before breakage. Brittle cracks are, on the other hand, events, which have no or relatively small amount of plastic history. The mode of crack growth is also different in ductile and brittle materials. In ductile fracture, the crack grows slowly and is accompanied by a large amount of plastic deformation. The existing crack will usually not extend unless an increased stress is applied. On the other hand, in brittle fracture, cracks spread very rapidly with little or no plastic deformation. Once they are initiated, the cracks in a brittle material not only continue to grow, but also increase in magnitude and even propagate with high speeds amounting to the speed of sound.

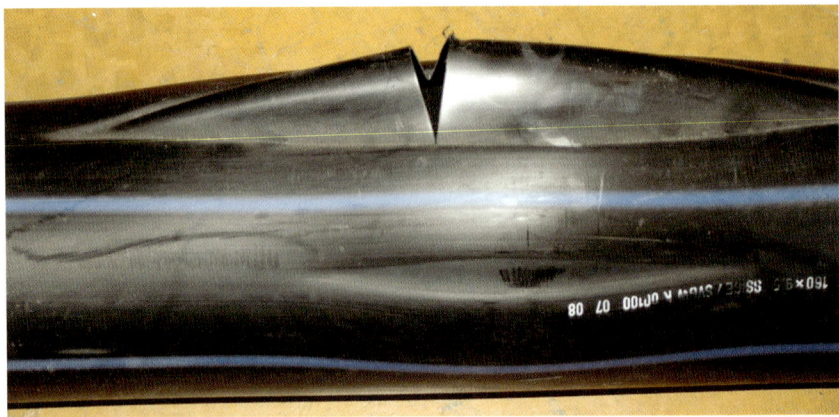

Fig. 3.2. A ductile crack in a PE-HD pipe under internal hydrostatic pressure.

On both macroscopic and microscopic levels, ductile fracture surfaces have larger necking regions and an overall rougher appearance than a brittle fracture surface. The surface consists of many micro-voids and dimples. One of the characteristics of ductile failure is that it occurs in the course of time and not instantaneously. On the contrary, brittle failure is a sudden effect with much shorter occurrence time.

The ductile fracture occurs in several stages: (1) First, the fibrils are stretched and in the interior of the material small micro-voids form. (2) The elastic–plastic deformation continues and the micro-voids enlarge to form a crack. (3) The crack continues to grow and it spreads in various directions. (4) Finally, crack propagation along a surface that inclined with the direction of the tensile stress axis takes place. Fig. 3.2 shows a ductile crack in a high-density polyethylene (PE-HD) pressure pipe under internal hydrostatic pressure. It shows the *ballooning* and *thinning* of the pipe wall at certain location. The ductile rupture occurs in this region. As shown in Fig. 3.2, the rupture cracks are normally in the transverse direction.

3.4   Fracture due to normal stress

Fracture of polymers due to normal stresses (i.e., the fracture mode I) can be divided into two main categories: (1) ductile fracture and (2) brittle fracture. The brittle fracture is identified by lack of plastic deformation. On the contrary, the ductile mode of fracture is preceded by stretching of fibrils. At the limit of deformation fracture occurs and the fibrils are detached from each other. After detachment, part of the deformation is recovered. Hence the fibrils spring back to some extent. Stretching and detachment of fibrils occur in the direction of normal force. Therefore, on the fracture surface, normal to the applied stress traces of stretching and spring-back phenomenon can be identified. Depending on the remaining length of the fibrils one may conclude whether the fracture was ductile or brittle. A remaining fibril length of about 1 μm is normally considered as a brittle mode of fracture. The topographical map of a ductile type of failure is like a relatively smooth hill view. The fracture surface of the brittle type, on the other hand, looks like sharp cliffs with sharp angles.

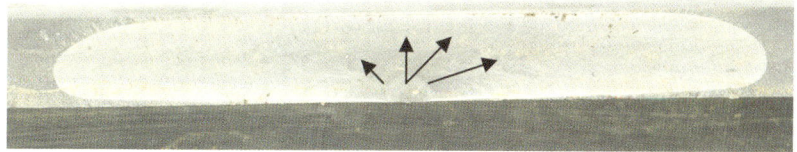

Fig. 3.3. Brittle crack growth in the pipe wall thickness of a PVC pipe. The bottom of the figure is the inside and the top is the outside of the pipe. The files show the direction of crack growth. Prior to sectioning of the wholly cracked pipe, white region indicating crazing was visible at the outer surface of the pipe. Beyond the white ellipsoidal region in the pipe wall, shown in the figure, a fast-moving crack in both directions along the pipe took place (see failure case study F-1 in this chapter).

3.5   Macroscopic observations on the fracture surfaces

Polymeric substances including thermoplastic, duroplastic, and elastomeric materials are distinguished by different deformation and fracture behavior:

(1)   The *thermoplastic* materials consist of crooked spaghetti-type linear molecular chains. Under tensile forces, these components are stretched and open up and become aligned along the tensile force. During this deformation process the molecular chains slide over each other. The deformation of these materials is time and temperature dependent; it is also dependent on the speed of loading. Due to time and temperature dependence, thermoplastic materials have relatively pronounced creep behavior. The fracture behavior of these materials at normal temperature is of the form of ductile creep rupture. At lower temperatures and higher loading speed, however, these materials can exhibit a pronounced brittle behavior. RCP in thermoplastic materials is a manifestation of this behavior.

(2)   The *duroplastic* materials have cross-links and their molecules are not totally free to expand. Therefore, overall deformation capacity of duroplastic materials is much less than thermoplastic polymers. These materials undergo smaller creep deformation than thermoplastics, but exhibit higher elastic response. The strain capacity of these materials is also limited to few percents. The fracture mode of duroplastic materials is mostly of the brittle type.

(3)   The *elastomeric* materials have relatively weak cross-links (i.e., smaller amount of molecular side bonds). At normal temperature, elastomers may have strains up to several hundred percents. Rupture of elastomers is due to the breakage of molecular chains.

For fracture analysis of pipes, the nature of the crack, the region of crack initiation, the mode of crack growth, and the extent of cracking are important. A crack may be initiated by crazing in a certain zone, which may be developed into a fracture. The fracture may remain stationary or may grow slowly or it may propagate at high speed. Fig. 3.3 shows the crack growth in the pipe wall thickness of a rigid polyvinyl chloride (PVC-H) buried pressure pipe.

3.6   Crack types

Cracks in polymer materials may be classified into various groups. A general classification contains the cracks due to various agents and cracks with various rate of growth. For plastic materials it is useful to identify various categories of cracks, these are:

(1) Environmental stress cracking
Environmental stress cracking (ESC) is a phenomenon which occurs under the simultaneous action of stress/strain and fluidic medium, in liquid or gaseous state. The result of ESC is the *crazing* in the plastic material. Stress cracking (*crazing*) is not the same as the *stress corrosion*. Crazing is related to the action of chemicals in a stressed plastic. The action of chemicals can eventually lead to the degradation of chemical bonds of the polymer. The ESC may have regular shape; it may also be quite irregular. For example, a bent thermoplastic pipe in some chemical environment may experience ring cracking. On the other hand, the cracks on the surface of a pipe with initial stresses may be quite irregular. The ESC is characterized by the embitterment of the plastic material.

(2) Thermal stress cracking
Thermal stress cracking of plastics occurs mostly when a plastic is exposed to constant stress for a longer time exposed to relatively high temperatures.

(3) Cracks due to long-term loading at various temperature
This type of cracking, also referred to as the creep rupture, occurs in the pipes under internal pressure at certain temperature.

(4) SCG
The so-called SCG is a phenomenon in which the state of an existing pre-crack or impurity may become unstable and a quasi-static fracturing would occur.

(5) RCP
RCP is a phenomenon characterized by a fast-moving crack along the pipe. Further aspects of this phenomenon shall be described in the following section.

3.7   RCP in plastic pipes

Rapid crack propagation (RCP) is a phenomenon in which a long fast-moving brittle crack can propagate in a material body. Cracking of glass plates and frozen lakes is an example of RCP. RCP can also occur in pies. Cast iron pipes and plastic pipes under certain conditions may also experience this phenomenon. In thermoplastic pipes fast-running cracks may propagate along a substantial length of the pipeline. Based on the results reported in the literature, the following general observations on RCP in PE pipes may be made:

(1) Under certain thermal and service conditions, including a loading state of internal pressure beyond a so-called *critical pressure*, fast-moving cracks having propagation speeds ranging from v = 70 to 370 m/s may occur. The critical pressure depends on temperature, pipe dimensions, material properties of the pipe, pipe processing, and residual stresses, ageing and service conditions, properties of the pressurizing fluid, supporting and/or embedding conditions, and finally, repair and maintenance conditions.

(2) It appears that above the critical pressure value, pseudo-steady-state crack propagation occurs. Below the critical pressure, the induced crack decelerates and, in a relatively short distance, comes to an arrest.

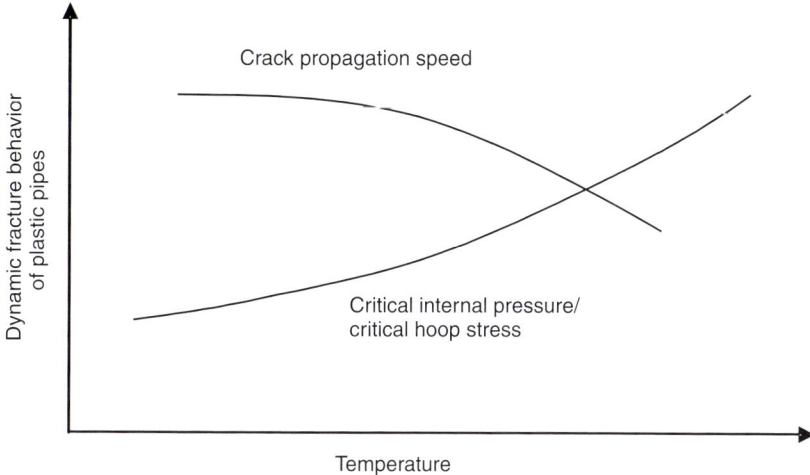

Fig. 3.4. Schematic graph of the influence of temperature on dynamic fracture of thermoplastic pipes and, in particular, on the crack speed, the critical stress, and the critical internal pressure.

(3) There is a so-called *critical thickness*, that is a certain thickness limit below which RCP cannot be sustained, while above that limiting value RCP may occur.

(4) The propagating crack runs swiftly along the pipe; generally, it traces a *curved* path. In some locations, under certain circumstances, the running crack may bifurcate and branch out into two or several paths; secondary branches may also be produced. Depending on the speed of propagation, the path of the propagating crack may be fairly straight or it may be strongly wavy of a generally sinusoidal form.

The influence of some of the above-mentioned parameters on dynamic cracking of PE pipes investigated by the S4 (Small Scale Steady State) laboratory tests is schematically represented in Fig. 3.4. This figure shows the deciding influence of temperature on the critical pressure, the critical stress, and the speed of crack propagation. In some reported field failures, the third-party damage had initiated circumferential RCP.

One of the motivations for the RCP study of polymer pipes arises from the need to assess the reliability of existing piping systems, some of which might have been operational for a number of years. Whether or not the ageing of the pipes might affect their dynamic fracture behavior is a question, which can only be answered through a systematic RCP research on the aged pipes.

An aspect of importance in dynamic behavior of polymer pipes is the influence of the production method. Depending on the production scheme, from a certain raw material various pipe elements with different dynamic fracture behavior may be produced. In this connection, the residual stress field caused by the practiced mode of manufacturing appears to be an influential factor. It is known that, in the process of extrusion, in the pipes which are cooled from outside a compressive hoop stress would be produced at the outer section of the pipe wall thickness while a tensile hoop stress would be produced in the inner part of the section. This residual stress field seems to strongly affect the dynamic fracture behavior of polymer pipes (Figs. 3.5–3.7).

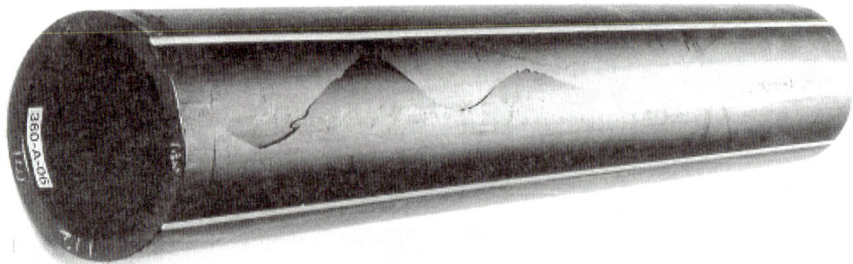

Fig. 3.5. Result of a S4 laboratory RCP test on a PE-HD pipe (Ø 225, standard diameter ratio (SDR) 11) at T = 0°C and 1.75 bar internal pressure. The figure shows the wavy path of crack along the pipe.

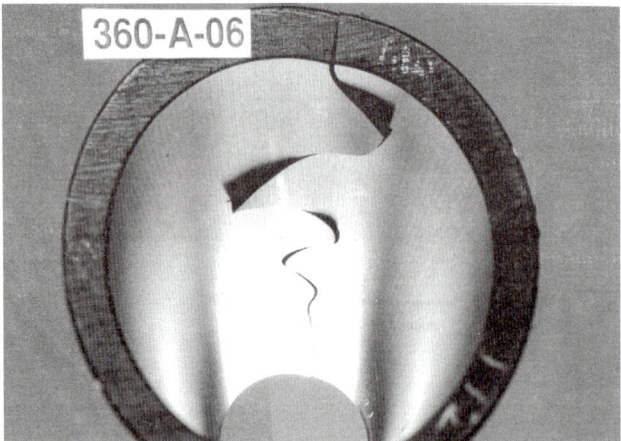

Fig. 3.6. Result of a S4 laboratory RCP test on a PE-HD pipe (Ø 225, standard diameter ratio (SDR) 11) at T = 0°C and 1.75 bar internal pressure. The figure shows the wavy path of crack along the pipe of Fig. 3.5 viewed from inside.

3.8   Types of fractures in pipes

Fracture assessment of plastic pipes may be based on various criteria. One of the convenient schemes for fracture investigation is the classification of cracks in plastic pipes in falling groups:

(1) Crazing
(2) Ring cracks
(3) Axial cracks
(4) Mixed cracking

Direction of crack propagation

Fig. 3.7. Result of a S4 laboratory RCP test on a PE-HD pipe (Ø 225, standard diameter ratio (SDR) 11) at T = 0°C and 1.75 bar internal pressure. The figure shows the crack front seen in the section through the pipe wall thickness.

Table 3.1 Classification of crazing in plastic pipes.

	Crazing											
	Through crazing				External crazing				Internal crazing			
Number	Single		Multiple		Single		Multiple		Single		Multiple	
Extent	Local	Global	Local	Global	Local	Global	Local	Global	Local	Global	Local	Global

Each of the above-mentioned types of cracks has its own features regarding the position in the pipe, dimensions of the crack, number of cracks, and the crack appearance. The features of these four groups of fractures are presented in Tables 3.1–3.4. These tables can be used as a basis for the failure assessment of plastic pipes. In each individual failure investigation, however, case-specific features should be considered as enhancement or modification of these basic classifications.

3.8.1   Types of crazing

Table 3.1 summarizes the various tapes of probable crazing in plastic pipes. Crazing may occur on the outer surface, on the inner surface, or in the pipe wall thickness. Crazing indicates some degree of micro-ductility. Therefore, in Table 3.1, the brittle designation of types of crazing is not emphasized.

Table 3.2 Classification of ring cracks in plastic pipes.

	Through crack								External crack								Internal crack							
	Ring cracking																							
Number	Single				Multiple				Single				Multiple				Single				Multiple			
Extent	Local		Global		Local		Global		Local		Global		Local		Global		Local		Global		Local		Global	
Crack surface appearance	Brittle	Ductile	Brittle	Ductile	Brittle	Ductile	Brittle	Ductile	Brittle	Ductile	Brittle	Ductile	Brittle	Ductile	Brittle	Ductile	Brittle	Ductile	Brittle	Ductile	Brittle	Ductile	Brittle	Ductile

Table 3.3 Classification of axial cracks in plastic pipes.

	Through crack								External crack								Internal crack							
	Axial cracking																							
Number	Single				Multiple				Single				Multiple				Single				Multiple			
Extent	Local		Single		Local		Global		Local		Global		Local		Global		Local		Global		Local		Global	
Crack surface appearance	Brittle	Ductile	Brittle	Ductile	Brittle	Ductile	Brittle	Ductile	Brittle	Ductile	Brittl	Ductile	Brittle	Ductile	Brittle	Ductile	Brittle	Ductile	Brittle	Ductile	Brittle	Ductile	Brittle	Ductile

3.8.2 Types of ductile cracks

Tables 3.2–3.4 summarize the various tapes of probable cracking in plastic pipes. Ring cracking may occur on the outer surface, on the inner surface, or it may influence the whole pipe wall thickness. The cracks may have ductile or brittle behavior. Table 3.5 shows the types of total fracture and rupture of the pipe. These cases are designated as the *pipe breakage*.

Table 3.4 Classification of mixed cracks in plastic pipes.

	Through crack				External crack				Internal crack			
Number	Single	Single	Multiple		Single		Multiple		Single		Multiple	
Extent	Local	Single	Local	Global	Local	Global	Local	Global	Local	Global	Local	Global
Crack surface appearance	Brittle / Ductil	Brittle / Ductile	Brittle / Ductile	Brittle / Ductile	Brittle / Ductile	Brittl / Ductile	Brittle / Ductile	Brittle / Ductile	Brittle / Ductile	Brittle / Ductile	Brittle / Ductile	Brittle / Ductile

Table 3.5 Categories of plastic pipes rupture.

Pipe break

	Pipe				Fitting	Pipe and Fitting
Number	Single		Multiple			
Extent	Local	Global	Local	Global		
Size and direction	Sweating, Leakage, Fracture, Burst, Explosion		Sweating, Leakage, Fracture, Burst, Explosion		Leakage, Fracture, Burst, Explosion	Sweating, Leakage, Fracture, Burst, Explosion, Axial separation, Bending separation, Shear-off

3.9 Guideline for diagnosis of fractured pipe

This section contains a basic set of guidelines for failure investigation of plastic pipes as related to the fracture assessment. Tables 3.6–3.10 relate the fracture types with the possible causes including mechanical, thermal, chemical, biological, and time factors. The tables are prepared in compact matrix forms. The rows of each matrix designate

Table 3.6 Assessment of crazing.

Specification	Crazing																							
	Through crazing								External crazing								Internal crazing							
Number	Single				Multiple				Single				Multiple				Single				Multiple			
Extent	Local		Global		Local		Global		Local		Global		Local		Global		Local		Global		Local		Global	
Appearance	Brittle	Ductile	Brittle	Ductile	Brittle	Ductile	Brittle	Ductile	Brittle	Ductile	Brittle	Ductile	Brittle	Ductile	Brittle	Ductile	Brittle	Ductile	Brittle	Ductile	Brittle	Ductile	Brittle	Ductile
Influence																								
Material	+	+	+	+	+	+	+	+	+	+	+	+	+	+	+	+	+	+	+	+	+	+	+	+
Mechanical		+		+		+		+		+		+		+		+		+		+		+		+
Internal pressure		+		+		+		+		+		+		+		+		+		+		+		+
External pressure																								
Axial tension		+		+		+		+		+		+		+		+		+		+		+		+
Axial compression																								
Bending		+		+		+		+		+		+		+		+		+		+		+		+
Traffic load																								
Settlement		+		+		+		+		+		+		+		+		+		+		+		+
Uplift		+		+		+		+		+		+		+		+		+		+		+		+
Production																								
Impact										+		+		+		+								
Vibration																								
Fatigue		+		+		+		+		+		+		+		+		+		+		+		+
Residual stresses		+		+		+		+		+		+		+		+		+		+		+		+
Other		+		+		+		+		+		+		+		+		+		+		+		+

Thermal											
High-temperature inside											
High-temperature outside											
UV radiation											
Fire											
Frost											
Other											
Chemical											
Water											
Oxygen	+	+	+	+	+	+	+	+	+	+	+
Acids	+	+	+	+	+	+	+	+	+	+	+
Alkalis	+	+	+	+	+	+	+	+	+	+	+
Solvents											
Oil	+	+	+	+	+	+	+	+	+	+	+
Benzene											
Other											
Service conditions											
Abrasion											
Interventions	+	−	+	+	+	+	+	+	+	+	+
Other											
Biological											
Microbes											
Animals											
Other											
Ageing factors											
Long-term effects	+	+	+	+	+	+	+	+	+	+	+

Table 3.7 Assessment of ring cracking.

Specification: Number	Through crack								External crack								Internal crack							
Extent	Single				Multiple				Single				Multiple				Single				Multiple			
Crack surface	Local		Global		Local		Global		Local		Global		Local		Global		Local		Global		Local		Global	
Crack surface	Brittle	Ductile	Brittle	Ductile	Brittle	Ductile	Brittle	Ductile	Brittle	Ductile	Brittle	Ductile	Brittle	Ductile	Brittle	Ductile	Brittle	Ductile	Brittle	Ductile	Brittle	Ductile	Brittle	Ductile
Influence																								
Material	+	+	+	+	+	+	+	+	+	+	+	+	+	+	+	+	+	+	+	+	+	+	+	+
Mechanical	+	+	+	+	+	+	+	+	+	+	+	+	+	+	+	+	+	+	+	+	+	+	+	+
Internal pressure	+	+	+	+	+	+	+	+	+	+	+	+	+	+	+	+	+	+	+	+	+	+	+	+
External pressure	+	+	+	+	+	+	+	+	+	+	+	+	+	+	+	+	+	+	+	+	+	+	+	+
Axial tension	+		+		+		+		+		+		+		+		+		+		+		+	
Axial compression																								
Bending	+	+	+	+	+	+	+	+	+	+	+	+	+	+	+	+	+	+	+	+	+	+	+	+
Traffic load	+	+	+	+	+	+	+	+	+	+	+	+	+	+	+	+	+	+	+	+	+	+	+	+
Settlement	+	+	+	+	+	+	+	+	+	+	+	+	+	+	+	+	+	+	+	+	+	+	+	+
Uplift	+	+	+	+	+	+	+	+	+	+	+	+	+	+	+	+	+	+	+	+	+	+	+	+
Production	+	+	+	+	+	+	+	+	+	+	+	+	+	+	+	+	+	+	+	+	+	+	+	+
Impact	+	+	+	+	+	+	+	+	+	+	+	+	+	+	+	+	+	+	+	+	+	+	+	+

Vibration	+	+	+	+	+	+	+	+	+	+	+	+	+	+	+	+	+	+	+	+	+	+
Fatigue	+	+	+	+	+	+	+	+	+	+	+	+	+	+	+	+	+	+	+	+	+	+
Residual stresses																						
Other																						
Thermal																						
High-temperature inside	+	+	+	+	+	+	+	+	+	+	+	+	+	+	+	+	+	+	+	+	+	+
High-temperature outside	+	+	+	+	+	+	+	+	+	+	+	+	+	+	+	+	+	+	+	+	+	+
UV radiation																						
Fire																						
Frost	+	+	+	+	+	+	+	+	+	+	+	+	+	+	+	+	+	+	+	+	+	+
Other																						
Chemical																						
Water																						
Oxygen																						
Acids																						
Alkalis																						
Solvents																						
Oil																						
Benzene																						
Other																						
Service conditions																						
Abrasion	+	+	+	+	+	+	+	+	+	+	+	+	+	+	+	+	+	+	+	+	+	+
Interventions	+	+	+	+	+	+	+	+	+	+	+	+	+	+	+	+	+	+	+	+	+	+
Other																						
Biological																						
Microbes																						
Animals																						
Other																						
Ageing factors																						
Long-term effects	+	+	+	+	+	+	+	+	+	+	+	+	+	+	+	+	+	+	+	+	+	+

Table 3.8 Assessment of axial cracking.

	Axial cracking																							
Specification	**Through crack**								**External crack**								**Internal crack**							
Number	Single				Multiple				Single				Multiple				Single				Multiple			
Extent	Local		Global		Local		Global		Local		Global		Local		Global		Local		Global		Local		Global	
Crack surface	Brittle	Ductile	Brittle	Ductile	Brittle	Ductile	Brittle	Ductile	Brittle	Ductile	Brittle	Ductile	Brittle	Ductile	Brittle	Ductile	Brittle	Ductile	Brittle	Ductile	Brittle	Ductile	Brittle	Ductile
Influence																								
Material	+	+	+	+	+	+	+	+	+	+	+	+	+	+	+	+	+	+	+	+	+	+	+	+
Mechanical	+	+	+	+	+	+	+	+	+	+	+	+	+	+	+	+	+	+	+	+	+	+	+	+
Internal pressure	+	+	+	+	+	+	+	+	+	+	+	+	+	+	+	+	+	+	+	+	+	+	+	+
External pressure																								
Axial tension																								
Axial compression																								
Bending	+	+	+	+	+	+	+	+	+	+	+	+	+	+	+	+	+	+	+	+	+	+	+	+
Traffic load	+	+	+	+	+	+	+	+	+	+	+	+	+	+	+	+	+	+	+	+	+	+	+	+
Settlement																								
Uplift																								
Production	+	+	+	+	+	+	+	+	+	+	+	+	+	+	+	+	+	+	+	+	+	+	+	+
Impact																								
Vibration																								

	1	2	3	4	5	6	7	8	9	10	11	12	13	14	15	16	17	18	19	20	21	22
Fatigue	+	+	+	+	+	+	+	+	+	+	+	+	+	+	+	+	+	+	+	+	+	+
Residual stresses	+	+	+	+	+	+	+	+	+	+	+	+	+	+	+	+	+	+	+	+	+	+
Other																						
Thermal																						
High-temperature inside																						
High-temperature outside																						
UV radiation																						
Fire																						
Frost	+	+	+	+	+	+	+	+	+	+	+	+	+	+	+	+	+	+	+	+	+	+
Other																						
Chemical																						
Water																						
Oxygen																						
Acids																						
Alkalis																						
Solvents																						
Oil																						
Benzene																						
Other	+	+	+	+	+	+	+	+	+	+	+	+	+	+	+	+	+	+	+	+	+	+
Service conditions																						
Abrasion																						
Interventions	+	+	+	+	+	+	+	+	+	+	+	+	+	+	+	+	+	+	+	+	+	+
Other																						
Biological																						
Microbes																						
Animals																						
Other																						
Ageing factors																						
Long-term effects	+	+	+	+	+	+	+	+	+	+	+	+	+	+	+	+	+	+	+	+	+	+

Table 3.9 Assessment of mixed cracking.

Specification	Through crack								External crack								Internal crack							
Number	Single				Multiple				Single				Multiple				Single				Multiple			
Extent	Local		Global		Local		Global		Local		Global		Local		Global		Local		Global		Local		Global	
Crack surface	Brittle	Ductile	Brittle	Ductile	Brittle	Ductile	Brittle	Ductile	Brittle	Ductile	Brittle	Ductile	Brittle	Ductile	Brittle	Ductile	Brittle	Ductile	Brittle	Ductile	Brittle	Ductile	Brittle	Ductile
Influence																								
Material	+	+	+	+	+	+	+	+	+	+	+	+	+	+	+	+	+	+	+	+	+	+	+	+
Mechanical																								
Internal pressure																								
External pressure																								
Axial tension																								
Axial compression																								
Bending																								
Traffic load																								
Settlement																								
Uplift																								
Production	+	+	+	+	+	+	+	+	+	+	+	+	+	+	+	+	+	+	+	+	+	+	+	+
Impact	+	+	+	+	+	+	+	+	+	+	+	+	+	+	+	+	+	+	+	+	+	+	+	+
Vibration	+	+	+	+	+	+	+	+	+	+	+	+	+	+	+	+	+	+	+	+	+	+	+	+

Fatigue	Residual stresses	Other	*Thermal*	High-temperature inside	High-temperature outside	UV radiation	Fire	Frost	Other	*Chemical*	Water	Oxygen	Acids	Alkalis	Solvents	Oil	Benzene	Other	*Service conditions*	Abrasion	Interventions	Other	*Biological*	Microbes	Animals	Other	*Ageing factors*	Long-term effects
+	+	+				+		+	+			+	+	+	+	+	+	+			+			+				+
+	+	+				+		+	+			+	+	+	+	+	+	+			+			+				+
+	+	+				+		+	+			+	+	+	+	+	+	+			+			+				+
+	+	+				+		+	+			+	+	+	+	+	+	+			+			+				+
+	+	+				+		+	+			+	+	+	+	+	+	+			+			+				+
+	+	+				+		+	+			+	+	+	+	+	+	+			+			+				+
+	+	+				+		+	+			+	+	+	+	+	+	+			+			+				+
+	+	+				+		+	+			+	+	+	+	+	+	+			+			+				+
+	+	+				+		+	+			+	+	+	+	+	+	+			+			+				+
+	+	+				+		+	+			+	+	+	+	+	+	+			+			+				+
+	+	+				+		+	+			+	+	+	+	+	+	+			+			+				+
+	+	+				+		+	+			+	+	+	+	+	+	+			+			+				+
+	+	+				+		+	+			+	+	+	+	+	+	+			+			+				+
+	+	+				+		+	+			+	+	+	+	+	+	+			+			+				+
+	+	+				+		+	+			+	+	+	+	+	+	+			+			+				+
+	+	+				+		+	+			+	+	+	+	+	+	+			+			+				+
+	+	+				+		+	+			+	+	+	+	+	+	+			+			+				+
+	+	+				+		+	+			+	+	+	+	+	+	+			+			+				+
+	+	+				+		+	+			+	+	+	+	+	+	+			+			+				+
+	+	+				+		+	+			+	+	+	+	+	+	+			+			+				+

Table 3.10 Assessment of plastic pipe rupture.

Pipeline element	Pipe										Fitting				Pipe and fitting							
Number	Single					Multiple																
Extent	Local		Global			Local		Global														
Size and direction	Sweating	Leakage	Crack	Burst	Explosion	Sweating	Leakage	Crack	Burst	Explosion	Leakage	Crack	Burst	Explosion	Sweating	Leakage	Crack	Burst	Explosion	Displacement	Separation	Bending
Influence																						
Material	+	+	+	+	+	+	+	+	+	+	+	+	+	+	+	+	+	+	+	+	+	+
Mechanical	+	+	+	+	+	+	+	+	+	+	+	+	+	+	+	+	+	+	+	+	+	+
Internal pressure	+	+	+	+	+	+	+	+	+	+	+	+	+	+	+	+	+	+	+	+	+	+
External pressure	+	+	+	+	+	+	+	+	+	+	+	+	+	+	+	+	+	+	+	+	+	+
Axial tension	+	+	+	+	+	+	+	+	+	+	+	+	+	+	+	+	+	+	+	+	+	+
Axial compression	+	+	+	+	+	+	+	+	+	+	+	+	+	+	+	+	+	+	+	+	+	+
Bending	+	+	+	+	+	+	+	+	+	+	+	+	+	+	+	+	+	+	+	+	+	+
Traffic load	+	+	+	+	+	+	+	+	+	+	+	+	+	+	+	+	+	+	+	+	+	+
Settlement	+	+	+	+	+	+	+	+	+	+	+	+	+	+	+	+	+	+	+	+	+	+
Uplift	+	+	+	+	+	+	+	+	+	+	+	+	+	+	+	+	+	+	+	+	+	+
Production	+	+	+	+	+	+	+	+	+	+	+	+	+	+	+	+	+	+	+	+	+	+
Impact	+	+	+	+	+	+	+	+	+	+	+	+	+	+	+	+	+	+	+	+	+	+
Vibration	+	+	+	+	+	+	+	+	+	+	+	+	+	+	+	+	+	+	+	+	+	+

Leakage/pipe break

Fatigue	Residual stresses	Other	*Thermal*	High-temperature inside	High-temperature outside	UV radiation	Fire	Frost	Other	*Chemical*	Water	Oxygen	Acids	Alkalis	Solvents	Oil	Benzene	Other	*Service conditions*	Abrasion	Interventions	Other	*Biological*	Microbes	Animals	Other	*Ageing factors*	Long-term effects
+	+						+	+	+						+					+	+	+		+	+		+	+
+	+						+	+	+						+					+	+	+		+	+		+	+
+	+						+	+	+						+					+	+	+		+	+		+	+
+	+						+	+	+						+					+	+	+		+	+		+	+
+	+						+	+	+						+					+	+	+		+	+		+	+
+	+						+	+	+						+					+	+	+		+	+		+	+
+	+						+	+	+						+					+	+	+		+	+		+	+
+	+						+	+	+						+					+	+	+		+	+		+	+
+	+						+	+	+						+					+	+	+		+	+		+	+
+	+						+	+	+						+					+	+	+		+	+		+	+
+	+						+	+	+						+					+	+	+		+	+		+	+
+	+						+	+	+						+					+	+	+		+	+		+	+
+	+						+	+	+						+					+	+	+		+	+		+	+
+	+						+	+	+						+					+	+	+		+	+		+	+
+	+						+	+	+						+					+	+	+		+	+		+	+
+	+						+	+	+						+					+	+	+		+	+		+	+
+	+						+	+	+						+					+	+	+		+	+		+	+
+	+						+	+	+						+					+	+	+		+	+		+	+
+	+						+	+	+						+					+	+	+		+	+		+	+
+	+						+	+	+						+					+	+	+		+	+		+	+
+	+						+	+	+						+					+	+	+		+	+		+	+
+	+						+	+	+						+					+	+	+		+	+		+	+
+	+						+	+	+						+					+	+	+		+	+		+	+

various possible environmental effects and the columns depict the types of fracture, which may occur in the pipe. The probability of an environmental factor being responsible for a specific fracture event is designated by the "+" sign in the matrix. The empty elements of the tables are indications that no general correlation may exist between the environmental causes and the fracture events. However, in a specific case, certain correlation (or lack of correlation) may be assesses.

It should be emphasized that the entries in Tables 3.6–3.10 are based on the knowledge of the author and the inputs from some other experts and the experience gained from various failure investigations. These assessments should be regarded only as general guidelines on which the individual assessments with case-specific modifications and judgment can be based. These tables can be used for failure assessments; they also can be used as a *knowledge base* for creation of an *expert system* for failure diagnosis of plastic pipes.

3.10   Failure case studies

In this section, a number of case studies related to fracture of plastic pipes are presented. As in the other chapters, a unified compact tabular form is adopted for each case. The tabular outline is followed by a number of sketches and photographs. Each of these cases was supplemented by a comprehensive documentation containing the details of the case. However, for brevity of presentation, in the outline of these failure cases the details are not included. Moreover, the information and proofs leading to the hypotheses and the failure judgments are not elucidated.

It is to be emphasized that the failure cases outlined in this section are statistical events, which have been caused by specific circumstances. The fact that a certain material or pipe type did fail would by no means reflect the weakness or malfunction related to that material. The circumstances leading to these events could have also caused damages in other pipe materials and products as well. One should bear in mind that the material and pipe are only two factors among the many that could cause a failure to occur. Hence the cases outlined in this section should be treated in an educational and supportive perspective and should be devoid from any value judgment and generalization about the specific materials and products.

The list of failure cases presented is as follows:

Case F-1:   Failure of PVC water pressure pipeline (Table 3.11)
Case F-2:   Cracks in a PVC pipe (Table 3.12)
Case F-3:   Rupture of glass-fiber-reinforced polyester (GFRP) pressure pipeline (Table 3.13)
Case F-4:   Cracks in the GFRP pipe (Table 3.14)
Case F-5:   Failure due to internal hydrostatic pressure testing (Table 3.15)
Case F-6:   Failure of GFRP pipeline (Table 3.16)
Case F-7:   Failure of GFRP pipeline (Table 3.17)
Case F-8:   Cracking of a PE T joint (Table 3.18)
Case F-9:   Breakage of sea water thermoplastic pipeline (Table 3.19)

Table 3.11 Case study of a PVC drink water supply pipe.

Failure case F-1	Failure of PVC water pressure pipeline
Pipe material and dimensions	PVC-U (nominal diameter (DN) = 225 mm; wall thickness, 16.7 mm; pipe type nominal pressure (PN) = 16)
Description of the system	Buried pipeline in an agricultural region; trench depth over the pipe crown, 1.08 m; trench width, 0.8 m; length, 3 km. A reservoir was placed at the altitude of 600 m over the sea level. The altitude of the failure place was 470 m and the distance from the reservoir to the failure place was 5000 m Service pressure: ca. 13.4 bar Service conditions: tractors and possibility of service traffic
Failed part	Rupture of two pipes during several months
Observed phenomenon	RCP Longitudinal and helical cracks on two pipes
Failure description	Two cases of pipe failure were investigated: Both pipes had through cracks that run along the whole pipe. The cracks in one of the two pipes had wavy path and other pipe was broken through relatively straight cracking. Both types of cracks showed a typical picture of RCP. The crack surfaces were brittle
Environmental conditions	Internal hydrostatic pressure, soil pressure, occasional service traffic load, possibility of water hammer, point loads due to large gravels and remaining installation wood pieces under the pipe
Time to failure	Four failures within 3 years after 11 years service life
Tests performed	Long-term hydrostatic pressure tests, long-term hydrostatic pressure tests with two diametrical external point force, torsion pendulum tests, material bending tests, split disk test, microscopic examination (the microscopic examination showed indication of material flaws)
Other investigations	Site investigation Large pieces of stones were found in the embedment Moreover, some forgotten pieces of transversely placed installation wood pieces were found under the excavated pipe
Failure cause(s)	Material flows, water hammer superposed on internal hydrostatic pressure and additional effects due to residual stresses
Suggested corrective actions	Improvement of material quality and embedment conditions
Photo documentation	See related photos and sketches (Figs. 3.8–3.15)

Fig. 3.8. Ruptured PVC-U pipe in trench. The backfilling has been removed.

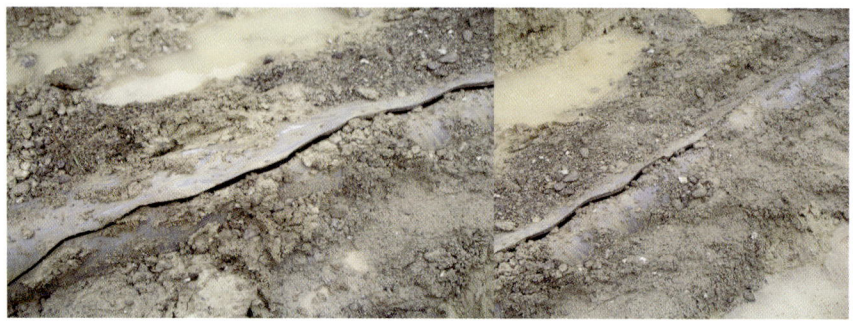

Fig. 3.9. Detailed photograph of the ruptured PVC pipe in trench:
The backfilling has been removed.

Fig. 3.10. Excavated broken pipe on-site. The figure shows the pipe pieces and socket at one end.

Fig. 3.11. Assembled pieces of the broken PVC pipe. Some pieces have been cut out for material tests.

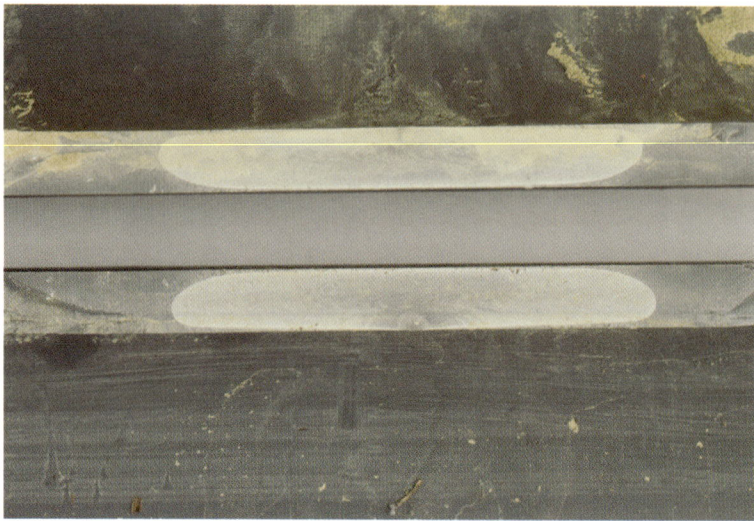

Fig. 3.12. Section through the pipe wall thickness in the zone of possible crack initiation at both sides of the wall thickness this picture shows the white zone and the material imperfection (possibly an inclusion) core from which the crack was initiated. Pipe wall thickness was 16.7 mm and the length of white zone was about 150 mm.

Fig. 3.13. Section through the pipe wall thickness in the zone of possible crack initiation at one side of the wall thickness this picture shows the white zone and the material imperfection (possibly an inclusion) core from which the crack was initiated. Pipe wall thickness was 16.7 mm and the length of white zone was about 150 mm.

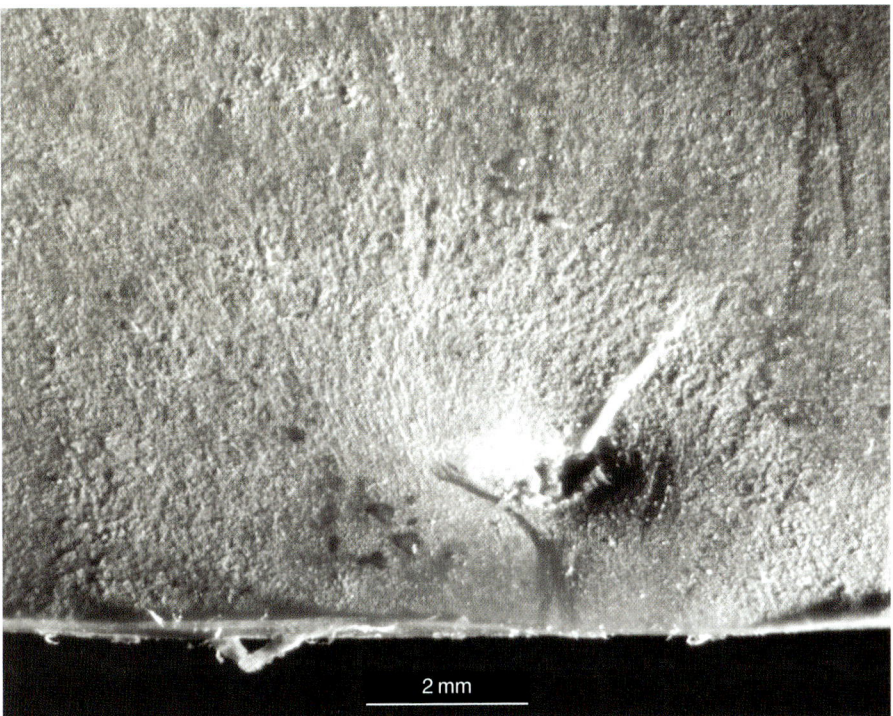

Fig. 3.14. Microscopic picture of the crack initiation zone in the pipe wall thickness. The figure shows an imperfection in the inner part of the pipe wall thickness.

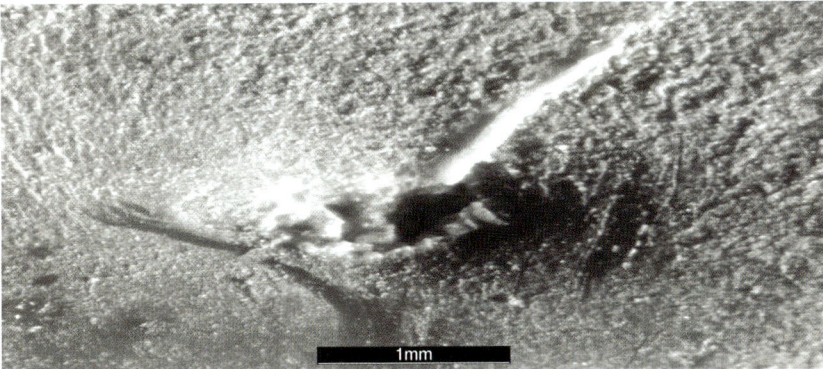

Fig. 3.15. The more detailed microscopic picture of the material imperfection core (inclusion) in the pipe wall thickness.

Table 3.12 Case study of cracking of a PVC pipe.

Failure case F-2	Cracks in a PVC pipe
Pipe material and dimensions	PVC-C pipe (nominal pressure (PN) = 6 bar, nominal diameter (DN) = 160 mm)
Description of the system	PVC pressure pipeline in collection canal, partly embedded in concrete
Failed part	Pipe
Observed phenomenon	Fracture in the non-embedded exposed part
Failure description	Fracture surface: glassy and brittle
Environmental conditions	Service pump pressure: maximum 5.2 bar with possibility of negative internal pressure
Time to failure	16 years in service
Tests performed	Material tests
Other investigations	Site investigation
Failure cause(s)	Inappropriate bedding, overloading through pump action, increase beyond allowable stress
Suggested corrective actions	Improving the bedding conditions
Photo documentation	See Fig. 3.16

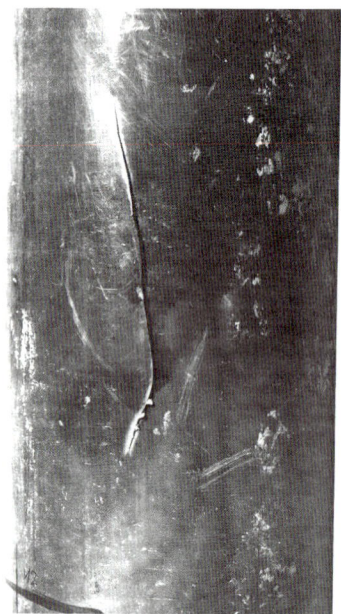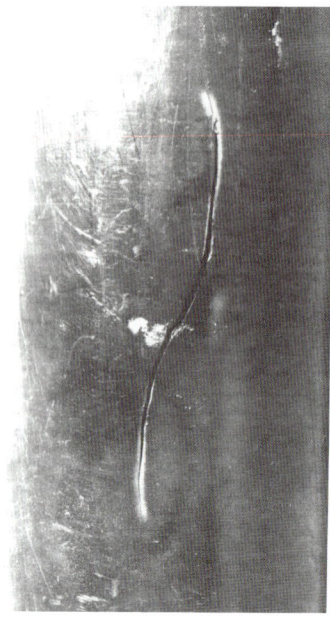

Fig. 3.16. Crack on the PVC pipe.

Table 3.13 Case study of a water turbine GFRP pipe.

Failure case F-3	Rupture of GFRP water turbine pressure pipeline
Pipe material and dimensions	GFRP (GF-UP) pipe (nominal diameter (DN) = 1400 mm; wall thickness about 33 mm; nominal pressure (PN) = 12 bar)
Description of the system	GFRP pressure buried pipeline for water turbine, located in a forest region under a service road; backfilling height about 700 mm
Failed part	Failure in the pipe and in the flange connection
Observed phenomenon	Total rupture and delamination of a pipe piece from the pipeline at the crown region (an angle of about 120°) affecting the pipe and the end ring connection
Failure description	Extensive cracking and delamination and fragmentation of some parts at the crown region. The pipe wall section was made of a sandwich construction produced by centrifugal casting. The middle part of the pipe wall section consisted of sand with smaller amount of polyester as the internal and external layers. Delamination occurred mainly at the interface of these layers
Environmental conditions	Internal pressure and external effects acting on a buried pipeline; earth filling containing coarse stone aggregates
Time to failure	Immediately after installation during acceptance stage
Tests performed	Material tests, internal pressure test on an intact part of the pipe piece
Other investigations	Site investigation
Failure cause(s)	Material weakness, production insufficiencies, lack of engineering calculations, inappropriate bedding conditions, external loading. Hence, the material composition and the production features were mainly responsible for the failure of the pipe, which did not have enough reserve strength
Suggested corrective actions	Material improvement, appropriate engineering design, improvement of installation conditions
Photo documentation	See related photos (Figs. 3.17–3.21)

Fig. 3.17. The picture of the failure pipe showing the fragmented upper part with delaminations.

Fig. 3.18. The end view of the failed pipe at the flange region showing the delaminations.

Fig. 3.19. Detailed view of the delaminations seen from the end region.

Fig. 3.20. The upper part of the delaminated section.

Fig. 3.21. Internal hydrostatic pressure test on the intact part of the GFRP pipe (diameter 1400 mm).

Table 3.14 Cracks in a GFRP pipe to be used for tunnel drainage.

Failure case F-4	Cracks in a GFRP pipe to be used for tunnel drainage
Pipe material and dimensions	Glass-fiber-reinforced vinyl ester (GFRV) pipe (nominal diameter (DN) = 600 mm)
Description of the system	Drainage system to be embedded in concrete for drainage in a tunnel
Failed part	Pipe
Observed phenomenon	Irregular cracks including star-shaped cracks and longitudinal cracks at the internal surface of the part
Failure description	Extensive localized cracking in the coating at the internal surface, no sign of cracking at the external surface
Environmental conditions	Drainage ground water in tunnel, possible alkaline environment
Time to failure	Before installation or immediately after installation
Tests performed	Impact of falling bodies
Other investigations	Visual examination
Failure cause(s)	Impact of objects, falling of the pipe, damage during installation, insufficient bending stiffness, brittle behavior
Suggested corrective actions	Care in production, storage, transport, and installation, improvement of material
Photo documentation	See related photos (Figs. 3.22 and 3.23)

Fig. 3.22. End view of the GFRP pipe.

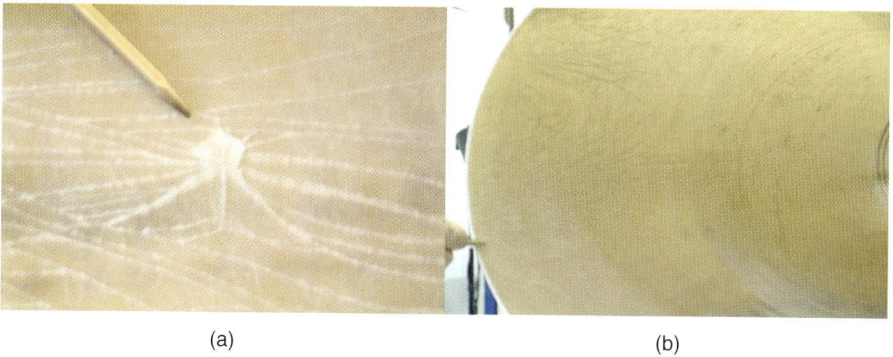

(a)   (b)

Fig. 3.23. (a) Cracks on the surface of the GRRP pipe and (b) cracks at the end zone of the pipe.

Table 3.15 Internal hydrostatic pressure tests related to the case of loss of tightness in an industrial drainage polypropylene pipe (see also the case in Chapter 5 under change of color).

Failure case F-5	Failure due to internal hydrostatic pressure testing
Pipe material and dimensions	Polypropylene
Description of the system	Pipe sample extracted from a polypropylene piping system
Failed part	Pipe sample
Observed phenomenon	Several longitudinal cracks
Failure description	The section through the pipe wall showed signs of thermal ageing
Environmental conditions	Long-term internal hydrostatic pressure test
Time to failure	Creep rupture
Tests performed	Internal hydrostatic pressure test at high temperature (T = 120°C)
Other investigations	Microscopic examination of the fracture surface; Site investigation of the piping system (see Chapter 5)
Failure cause(s)	Creep rupture at high temperature
Suggested corrective actions	Application temperature should be taken into account
Photo documentation	See Figs. 3.24 and 3.25 and also the figures of the related case in this chapter

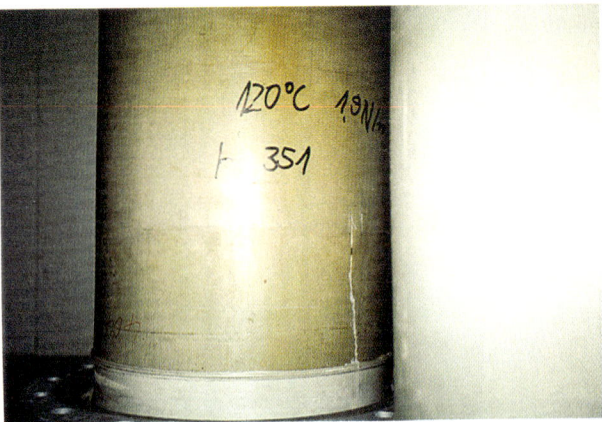

Fig. 3.24. Cracked portion of extracted polypropylene pipe sample in long-term internal pressure test at 80°C.

Fig. 3.25. Detained picture of the fractured zone in the pipe wall thickness of the pipe sample in Figure 3.24 showing thermal oxidation.

Table 3.16 Case of a ruptured GFRP pressure pipe.

Failure case F-6	Failure of GFRP pipeline
Pipe material and dimensions	GFRP pipe (diameter, 900 m)
Description of the system	GFRP pressure line – buried pressure pipeline for water power station located in an agricultural region at a depth of about 1 m
Failed part	Pipe and joint
Observed phenomenon	Pipe fracture at the crown region
Failure description	Totally ruptured pipe at the crown region and splitting of several pipe pieces with longitudinal as well as inclined through cracking
Environmental conditions	Internal water pressure, soil pressure, possibility of service traffic; in the failure zone several cable protection PE pipes crossed the GRFP pipe (see Fig. 3.26)
Time to failure	About 7 years
Tests performed	Bending and tensile tests on longitudinal and circumferential material samples
Other investigations	Site investigation
Failure cause(s)	A block consisting of four cable protection PE pipes ran parallel with along the main pipe. In a region of directional change, to reduce the curvature, the cable protection pipes crossed over the GFRP pipe. In this crossing, they were almost placed on the crown of the GFRP pipe. The line load of this system and the load due to the repair traffic caused the failure.

(*Continued*)

Table 3.16 (*Continued*)

Failure case F-6	Failure of GFRP pipeline
Suggested corrective actions	Improvement of material quality, avoidance of local concentrated loads such as large stones and, in this case, line loads from the cable protection pipes; dimensioning of pipelines for various critical loading conditions including the empty condition
Photo documentation	See Figs. 3.26–3.28

Fig. 3.26. Failure zone showing the main GFRP pipe and the crossing of cable protection pipe.

Fig. 3.27. Partial longitudinal view of the main GFRP pipe showing the fractured zone.

Fig. 3.28. Split zone at the crown of the pipe.

Table 3.17 Case of a GFRP pressure pipe failure.

Failure case F-7	Failure of GFRP pipeline
Pipe material and dimensions	GFRP pipe (diameter 800 mm; SN = 5000 N/m², nominal pressure (PN) = 10 bar)
Description of the system	Earth-embedded GFRP pressure pipeline for water power station located in a forest service load adjacent to a small river with no retaining wall at the riverside
Failed part	Pipe and joint
Observed phenomenon	Rupture of pipeline during control in empty condition
Failure description	Totally ruptured pipe at the bottom with longitudinal as well as inclined through cracking
Environmental conditions	Mountain water, soil pressure, possibility of service traffic
Time to failure	About 7 years; the failure occurred during a control period, in which the pipeline was in an empty condition
Tests performed	Bending and tensile tests on longitudinal and circumferential material samples, burst tests
Other investigations	Site investigation
Failure cause(s)	Lateral movement of the pipeline due to local movement of the bedding following heavy rainfall and lack of enough lateral support due to the proximity of the river and lack of retaining walls
Suggested corrective actions	Improvement of material quality, better embedment and filling material, provision of lateral support for the bedding. Dimensioning of pipelines for various critical loading conditions including the empty condition
Photo documentation	See the related photographs and sketches (Figs. 3.29–3.35)

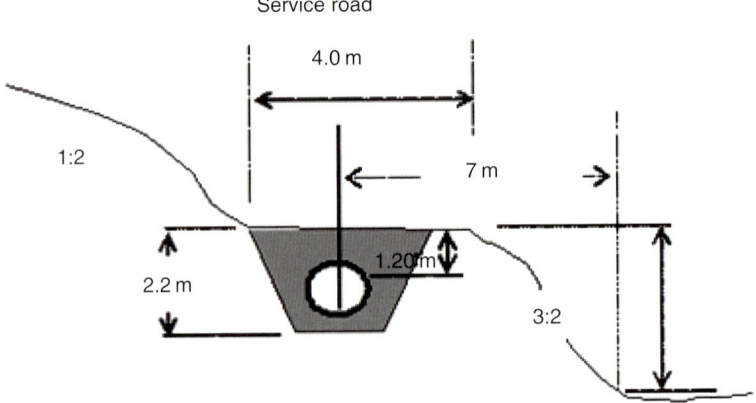

Fig. 3.29. Cross-section of the pipeline in the vicinity of the failure zone. The section is seen from the downstream side. The river is located at the right-hand side of this sketch.

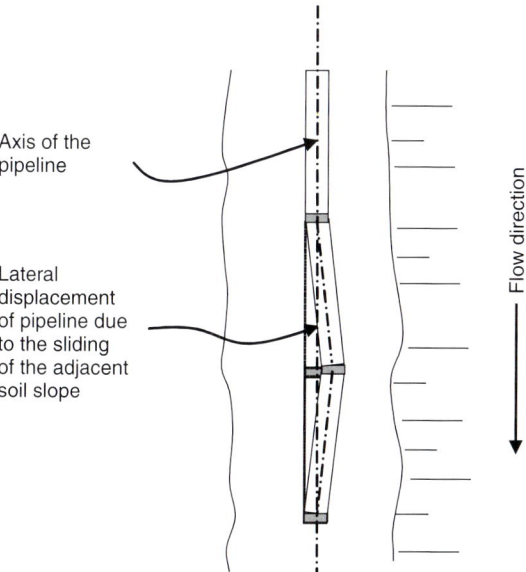

Axis of the pipeline

Lateral displacement of pipeline due to the sliding of the adjacent soil slope

Flow direction

Fig. 3.30. Schematics of the top view of the pipe system at the failure zone.

Fig. 3.31. A picture of the pipeline in the failure zone as viewed from downstream side. The river is located at the right-hand side of the picture. In the middle is the failed GFRP pipe near a cuff connection. At the left side is a PE pipe and at the right-hand side of the picture are two cable protection PVC pipes.

Fig. 3.32. Pipeline in the failure zone viewed from the downstream side; river is at the right-hand side.

Fig. 3.33. A detailed view of the failure zone. The PE pipe and the cable protection pipes are also seen at both sides of the failed pipe.

Fig. 3.34. Detailed view of the ruptured pipe. The figure shows the cracking at the bottom zone and the bedding condition; part of the stones might have been washed from the sides; the others were already present under the pipeline.

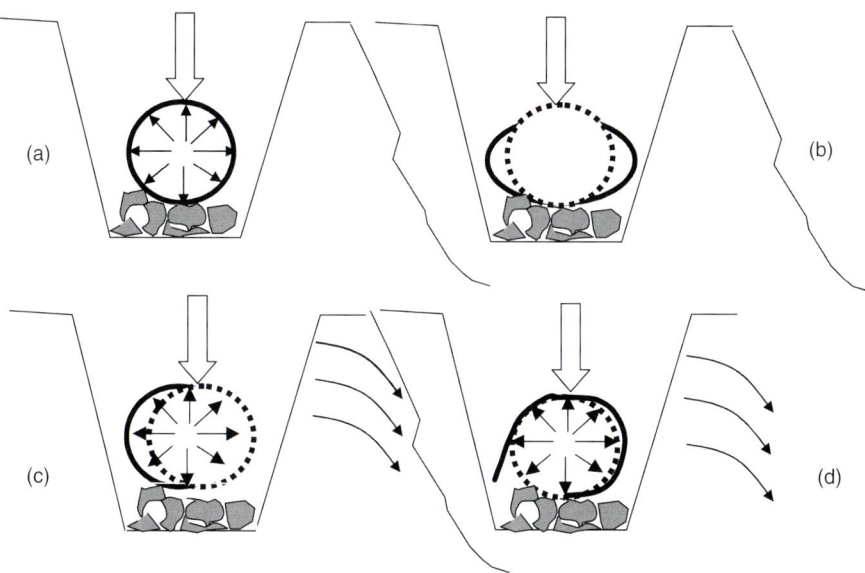

Fig. 3.35. Failure hypothesis: stages leading to failure of the pipeline. The figures show the cross-sectional sketch of the pipeline at the failure zone. (a) Pipe on coarse bedding with large pieces of stones and vertical soil pressure and point loads from the underlying stones. (b) Deformation of pipe under point load and vertical soil pressure. (c) Lateral displacement of the unprotected soil slope adjacent to the river. (d) Rupture of the pipe at the bottom zone due to combined action of vertical load, point load, frictional forces due to lateral land displacement.

Table 3.18 Case of a leakage of a PE-HD T joint in a drink water main.

Failure case F-8	Cracking of a PE T joint
Pipe material and dimensions	PE-HD, PE80 (nominal diameter (DN) = 315 mm)
Description of the system	Main water distribution installed in an installation tunnel. The piping system was installed on the ceiling of the tunnel. It was supported by metal supports at intervals of about 3 m apart
Failed part	PE T joint
Observed phenomenon	Cracking of one T joint and leakage
Failure description	Two longitudinal cracks, each about 70 mm long on the surface of the T joint
Environmental conditions	Internal: drink water; external: air with relatively constant temperature of about 15°C
Time to failure	Unknown
Tests performed	Computer tomography of the cracked T joint. The computer tomography revealed the presence of a hole with dimensions of about 30 mm × 10 mm inside of the wall thickness of the T joint. This hole was related to the observed surface cracks with small canals
Other investigations	Site investigation. After detecting the hole in one of the T joints an extensive examination of the whole pipeline was carried out. In this action, to investigate the possibility of existence of internal holes in other T joints and also in the pipes, about 6000 ultrasonic tests on the whole piping system was carried out. No other similar case was found
Failure cause(s)	T joint was produced by an injection molding technique. Further examination of the T section by computer tomography and sectioning revealed a large hole in the wall thickness, which was directly connected to the external cracks
Suggested corrective actions	Better injection molding technique evading the formation and trapping of internal air bubbles
Photo documentation	See Figs. 3.36–3.40

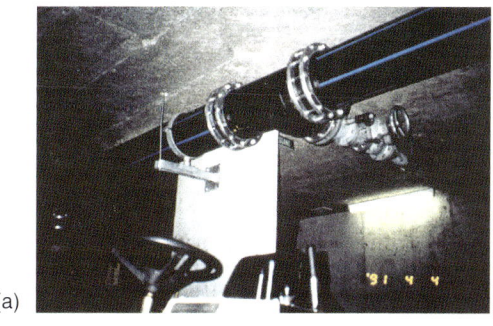

(a)

(b)

Fig. 3.36. (a) Part of the piping system and (b) PE T joint.

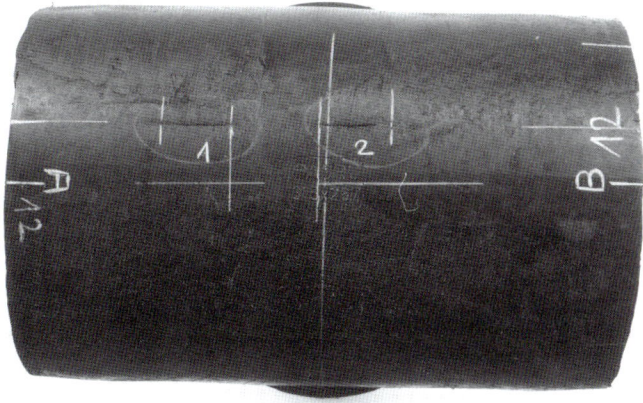

Fig. 3.37. Two surface cracks on the T joint.

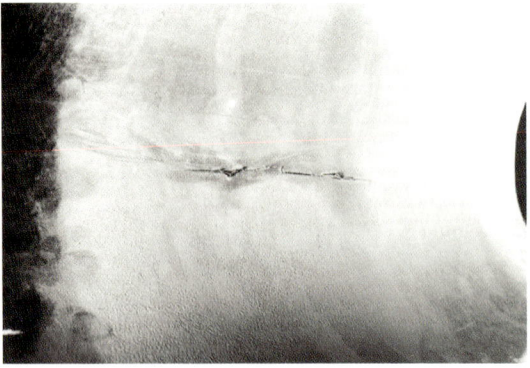

Fig. 3.38. Detailed view of one of the two surface cracks.

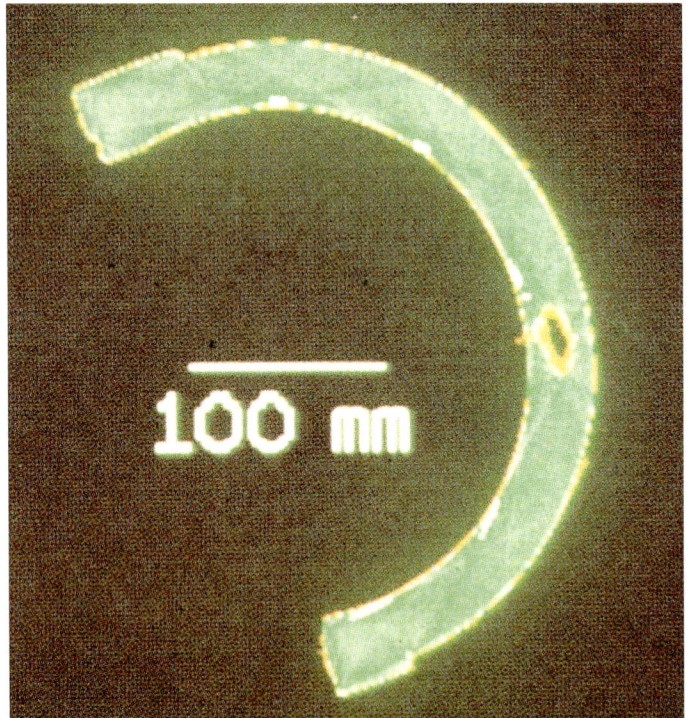

Fig. 3.39. Computer tomography picture of the section of the T joint showing a hole inside the wall thickness.

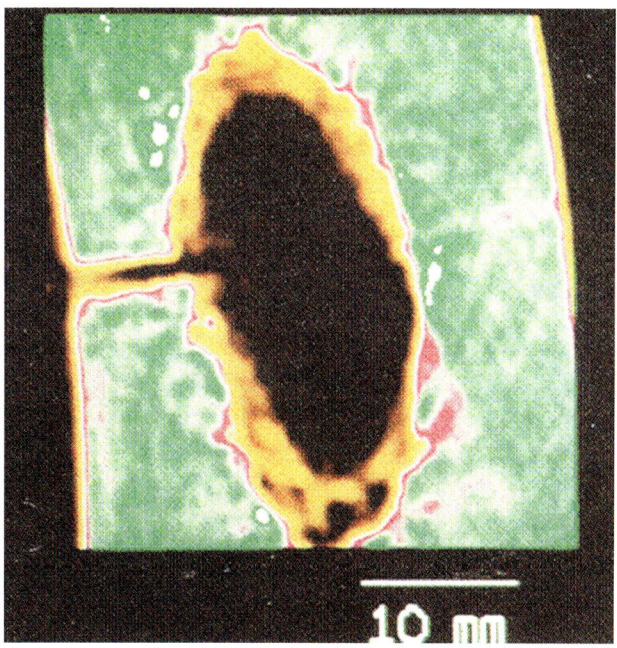

Fig. 3.40. A detailed computer tomography view of the hole inside of the wall thickness.

Table 3.19 Case of breakage of a submerged PE pipe.

Failure case F-9	Breakage of submerged thermoplastic pipeline.
Pipe material and dimensions	PE-HD (diameter 315 mm; wall thickness, 8 mm; pipe series, S20)
Description of the system	Water acquisition under pipeline about 10 m under the sea placed on the see floor and fixed by weights and transverse beams
Failed part	Pipe
Observed phenomenon	Breakage of the pipe, fragmentation into pieces
Failure description	Fragmented segments with distortion
Environmental conditions	Internal: fresh sea water; external: fresh sea water
Time to failure	About 6 years with a modification including addition fixation with transverse concrete elements
Tests performed	Unknown
Other investigations	Site investigation
Failure cause(s)	During the repair and modification work this location might have been impacted by the foreign objects. The cause of the failure can be classified under third-party damage "external intervention"
Suggested corrective actions	The damaged part was replaced. More care should be taken during the repair works.
Photo documentation	See Figs. 3.41–3.43

Fig. 3.41. Broken pieces of PE pipe.

Fig. 3.42. Positions of the rupture points (12 Uhr = 12 O'clock).

Fig. 3.43. Two fragments of the broken pipe.

3.11 Preventive measures against fracture of plastic pipes

Fracture of pipelines is one of the most frequent potential modes of failure. Fracture of a pipe system can have serious consequences. In sewer systems it may lead to contamination of the environment. In pressure water mains it may disrupt the water supply and cause side damages such as flooding of the streets and houses. In gas pipes, fracture may be the cause of explosions and more serious damages.

The failure cases encountered in practice show that fracture of plastic pipes may be traced back mainly to the following factors:

(1) Design and planning errors (insufficient dimensioning, lack of sufficient safety factors)
(2) Material weakness (material imperfections, brittle behavior, weak composite construction)
(3) Improper production method (residual stresses, hardening time)
(4) Initial damages (cracks, scratches)
(5) Installation errors (improper bedding conditions, concentrated forces)
(6) Weak joints
(7) Inappropriate service conditions
(8) Improper repairs
(9) External interventions

In the case of plastic pipe systems, factors such as thermal, chemical, and physical ageing generally cause material deterioration and embrittlement, and may contribute to potential fracture. The measures against fracture of plastic pipes should assure that the above-mentioned sources of error are eliminated. The bodies responsible for a pipe system should follow the whole service life of the pipe from the planning stage to its repair and replacement. For this purpose, appropriate quality control and health monitoring systems should be used. Furthermore, it is important to take into consideration the various load cases, applicability of the material, and the limits of the pipe service life. One should bear in mind that some of the second-order factors such as pieces of stones, installation wood pieces, and local damages may lead to fracture of a pipe. The best approach in arriving at preventive measures against potential is to plan the pipe system for the conceived applications, to choose the right material, to monitor the pipe system and to take the corrective measures during the service life of the pipeline.

4

Buckling of plastic pipes

4.1 Stability and instability

Stability of material systems including structures and piping systems is one of the major requirements. In simple terms, stability of a system implies that the system can maintain an existing state without major deviations from that state. The lack of stability of a system (i.e., the instability) may result changes in the properties and behavior of the system, which are not necessarily desired.

The structural systems including pipe systems may undergo certain deformations and stiffness changes that would reduce their strength and serviceability. Hence the instability is regarded as a failure mode, which should be seriously evaded. There are many cases in which structures have failed due to instability of their state of equilibrium. In buildings, vertical slender members may fail under relatively heavy loads; this type of instability is referred to as *buckling*. Buckling failure may occur in variety of cases, in compression members of trusses and frames, in piles, in pipes, in plate and shell structures, and in various machine elements. Pipes may also buckle laterally or longitudinally. The stability and instability of plastic pipes are strongly affected by the time and temperature as well as other system and environmental factors.

There are cases where the state of equilibrium of a structure may become dynamically unstable; this means that the structure may undergo vibratory motions with increasing amplitude. This type of instability has occurred in suspension bridges, in airplanes, in fluid conveying pipes, in rotating shafts, in impacted bodies, and in machine elements subjected to dynamic loading.

4.2 Modes of elastic instability

Elastic structures could loose their stability a number of ways. Modes of instability depend on the properties of the system itself and its environment including applied forces in the boundary conditions. Systems in static equilibrium may become statically and dynamically unstable. For structural systems, three types of instability may be identified; these are:

(1) Bifurcation of equilibrium – *classical buckling*
(2) Limit equilibrium instability – *snap-through buckling*
(3) Dynamic and *flutter* instabilities

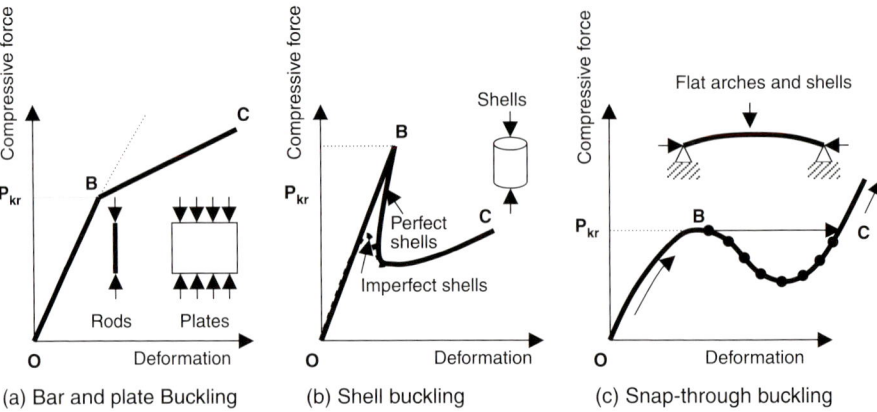

Fig. 4.1. Types of static elastic buckling.

4.2.1   Bifurcation of equilibrium

One of the salient features of static elastic instability is called the *bifurcation* of equilibrium state. At a certain stage of loading, the state of equilibrium of a structure may reach a point of bifurcation in which there are *two* possible paths (states) of equilibrium. The intersection of these two paths corresponds to the so-called *bifurcation of equilibrium*, because at such point two states of equilibrium can exist for the same load. Beyond the bifurcation point, the system can have one of the two choices of behavior. It can stay in its initial equilibrium regime or it could *diverge* from the *primary path* and follow a new path, the *secondary path*, of deformation. Fig. 4.1 schematically shows the bifurcation-type instability of an elastic system. From the physical point of view, the structure chooses the path corresponding to a minimum energy of the system.

The bifurcation point of an equilibrium state marks the *critical state* of behavior of an elastic system. The primary path (i.e., the initial state of equilibrium) beyond the bifurcation point is an unstable path while the secondary equilibrium path is stable. The loading condition corresponding to a bifurcation point is normally called the *critical load*.

4.2.2   Limit equilibrium instability

The loss of stability through the so-called *limitation of equilibrium* is characteristic of structures which carry the transverse loading mainly by compressive axial forces. Shallow arches and shallow shells are examples of such structures. In structures undergoing this type of instability there is no bifurcation point. The load–deformation curve of such systems is continuous and consists of a single curve with no branches; this curve has some stationary maximum and minimum points; the critical load corresponds to one of these maximal points.

A well-known type of *limitation of equilibrium instability* is referred to as the *snap-through* buckling. Snap-through buckling is a mode of instability in which an elastic system, under certain loading, may pass from an equilibrium state to a non-adjacent

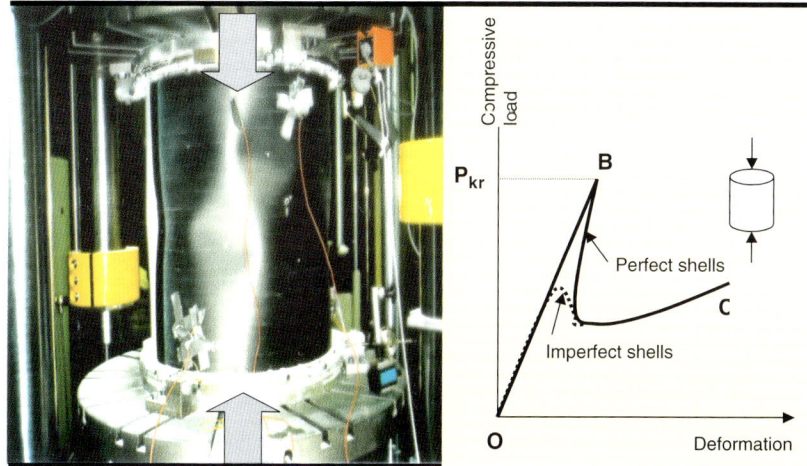

Fig. 4.2. Buckling and post-buckling behavior of a composite carbon-reinforced epoxy (CFRP) cylindrical shell under axial compression. The left figure shows the post-buckling mode of the CFRP cylinder in axial compression. The right picture shows the schematics of axial pressure as function of axial shortening of the cylinder.

equilibrium configuration. Fig. 4.2 schematically shows the snap-through type of insta-bility of an elastic system. The branch OB of the load–deformation curve describes the predominantly linear behavior of the arch. At the stationary point B, corresponding to a maximum applied force, the system "jumps" from a deformed state, marked by point B, to another deformation state much further away from its neighboring deformed configurations. Compressed shallow arches and shells can "snap-through" their bases and deform into reversed shapes undergoing *tensile* (instead of compressive) deforma-tions (Fig. 4.3).

4.2.3 Dynamic *and* flutter *instabilities*

Instability due to dynamic effects and the so-called *flutter* instability are both *dynamic instabilities*, in the sense that the motion and equilibrium of the system may become dynamically unstable; dynamical instability means that the unstable system would have *oscillations with increasing amplitude*. *Dynamic instability* of elastic systems can occur when the applied forces are *non-conservative*.

The *state of equilibrium* of an elastic system may also become unstable in a *dynamic fashion*. For example, an equilibrated structure under non-conservative forces may undergo oscillations with increasing amplitude. This is called *flutter instability*. Flutter instability can also occur in the structures having a steady motion; flutter of airplanes in motion is an important case of such phenomenon. In pressure pipe systems and also in underwater pipelines, flutter is a mode of increasing lateral vibrations. Under some conditions, flutter may cause additional stresses especially at the pipe joints.

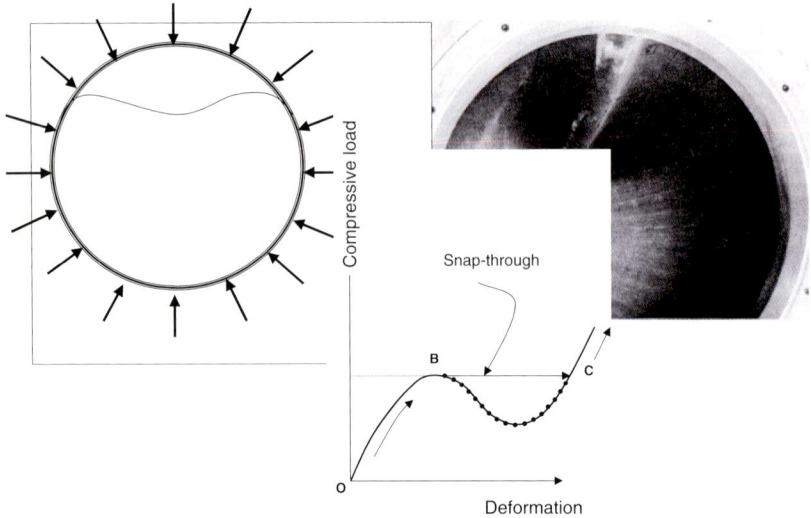

Fig. 4.3. Snap-through buckling of pipe under external hydrostatic pressure. The top left figure shows the snap-through phenomenon in a pipe; the top right figure shows the result of external hydrostatic pressure test on a polyvinyl chloride pipe demonstrating the snap-through phenomenon. The middle figure shows the lateral pressure versus shortening of a shallow shell and the schematics of the snap-through phenomenon.

The *state of motion* of a system under time-dependent forces can become dynamically unstable. For example, columns or struts under pulsating axial forces can loose their regular vibratory regime. Another example is the case of rotating shafts which at some value of rotational velocity may become dynamically unstable. Another example of dynamic instability is called *parametric resonance*. The parametric resonance is a mode of instability, which may occur under dynamic forces. In this type of instability, the mode of vibration of the system changes and an energy transfer takes place that could be detrimental for the systems. A bar subjected to dynamic loading would have axial vibrations, which under certain conditions may change to relatively wild lateral bending motions.

One special case of pipe buckling, which may occasionally occur in the submerged pipes, is known as the *propagating buckle*. In addition to various possible loadings, pipes placed on the seabed are additionally subjected to relatively high external hydrodynamic pressure. The external pressure may cause buckling in the pipe that may propagate along the pipe and affect a considerable length of the pipeline. To arrest the propagating buckle a number of the so-called *buckle stoppers* have been developed.

4.3   Excessive deformation

Excessive deformations are deformations, which are larger than the prescribed and calculated values. Deformations that may hinder the function of the pipe to expose it to collapse are regarded as excessive deformations. Excessive deformations may be in the form of large longitudinal deformations or flattening of the pipe cross-section. Excessive deformations are not necessarily the result of instability of pipes. For example, large partial settlement in a non-pressure gravity flow drainage pipe system may be regarded as excessive deformation, because it may hinder the free flow of the fluids or lead to settlement of solid materials in the pipe. Flattening of sewer pipes would reduce the hydraulic radius of the pipe and thus have negative effects on the fluid flow. From mechanical point of view, excessive deformations are highly nonlinear deformations, which go beyond the serviceability limits of the pipe.

4.4   General causes of buckling of plastic pipes

Buckling of plastic pipes is a mode of instability, which may be realized in several cases. Buckling of pipe can affect the service life of pipe system; the buckling event may also cause several side effects, including drastic financial consequences. This mode of instability may occur during production of pipe, storage and transport, installation, service conditions, and maintenance and repair. The main causes of buckling in pipe systems are:

(1) Wrong concept or incomplete planning
(2) Wrong design
(3) Pipe weakness (material and geometry)
(4) Defects and pre-deformations during manufacturing
(5) Damages and pre-deformations during storage, transport, and installation
(6) Environmental factors (mechanical loads, thermal and chemical effects)
(7) Damages and deformations during inspection and repair
(8) External intervention

Many standards and guidelines consider the buckling of pipe systems merely as the sectional buckling under lateral pressure. In fact piping systems may experience not only lateral, but also longitudinal buckling modes. Examples of potential longitudinal buckling events are exposed pipes at elevated temperature and highly inclined piping systems used in the inclined pipelines supplying the water turbines.

4.5   Buckling of buried pipes

Buried pipe systems are subjected to variety of environmental actions. The main loadings on buried pipes consist of internal pressure, earth pressure, traffic load, and external water pressure. Partial settlement, upheaval, pipe crossings, and ground motions are additional load cases that affect the pipe system behavior.

The buckling behavior of buried plastic pipes is affected by the interaction of the pipe and the surrounding soil. The relative stiffness of the pipe and the surrounding medium is an important factor determining the deformation mode of the pipe system. Fig. 4.4 schematically shows two modes of buckling behavior of buried pipe systems. One of these two modes is identified by the snap-through type of deformation, while the other mode is characterized by the ovalization of the pipe. The snap-through type of buckling occurs in buried pipes with stiff environment. The ovalization mode of deformation is mostly realizable in the cases in which the stiffness of the pipe is larger than the modulus of the side filling.

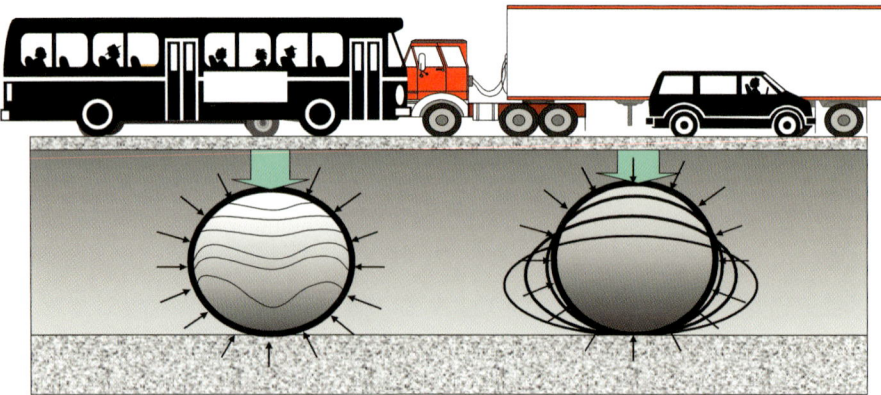

Fig. 4.4. Types of buckling in buried pipes. The schematic picture at the left shows the snap-through shape of instability. The schematic picture at the right shows the ovalization of the pipe under earth pressure and traffic load.

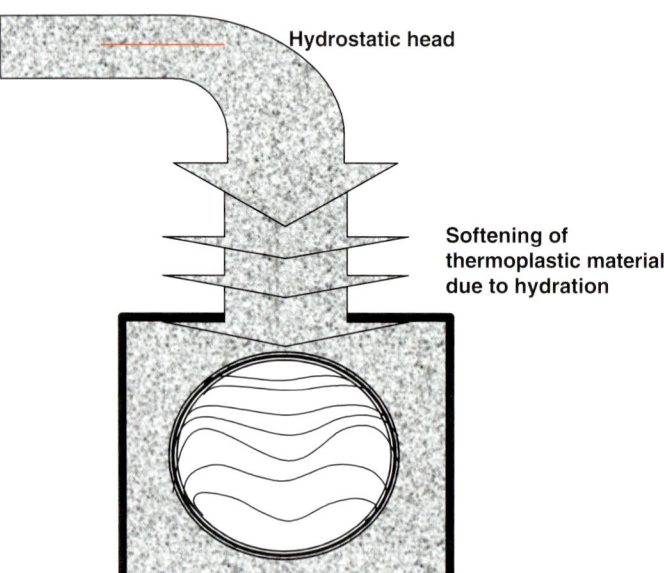

Fig. 4.5. Buckling process in a thermoplastic pipe during concreting.

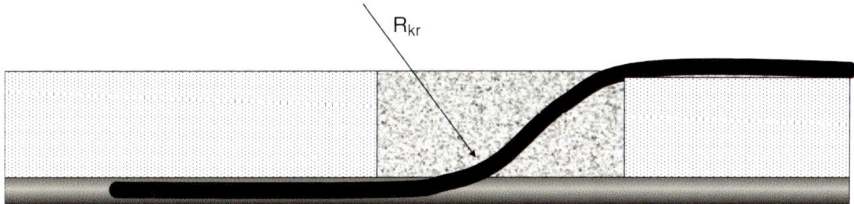

Fig. 4.6. Critical bent radius of a pipe; the pipe is drawn from the top of a trench into another pipe inside the trench.

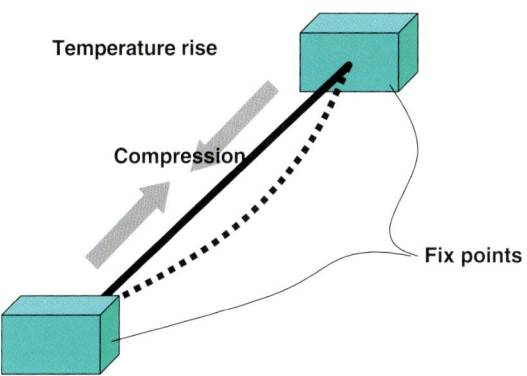

Fig. 4.7. Longitudinal buckling of a segment of a pipe system between two fixed points due to thermal increase.

4.6   Buckling during installation

Plastic pipes may buckle during handling and installation. One of the cases in which buckling may occur is shown in Fig. 4.5. This figure shows a continuous plastic pipe which is *pushed down* from the top of a trench into another pipe buried into a trench. If the bent radius is not large enough, then the buckling type known as the *Brazier effect* may occur. In this connection one may define and calculate a so-called bent *critical radius* below which the danger of buckling would exist (Fig. 4.6).

4.7   Buckling due to elevated temperatures

Exposed pipes may buckle at elevated temperatures. This phenomenon may occur in all pipe materials including plastics and metallic pipes. The pipe tends to undergo longitudinal expansion at increased temperature. If the expansion is hindered by some action or object such as fix points, then a longitudinal compression will be created. At high enough temperature gradient, the pipe may buckle under this compressive force. Fig. 4.7 schematically shows the longitudinal buckling of segment of a pipe system at elevated temperature between two fix points.

Another probable type of buckling of plastic pipes at high temperatures is the case of fire inside or outside of the pipe. The ignitable fluids and chemicals inside of the pipe

Fig. 4.8. Photographs showing the distorted shape of thermoplastic pipe samples in fire tests.

may cause a fire accident that may produce large deformations in the plastic pipes. Fig. 4.8 shows the result of a flame experiment on a polyethylene (PE) pipe. The flame was produced by a flame inside of the pipe.

4.8   Buckling of concrete-embedded pipes

In some applications, plastic pipes are partially or totally embedded inside of a concrete block. Typical applications are cable protection pipes, drainage pipe in tunnels, and water turbine inclined pipes. A wrong conception among some people is that the concrete-embedded pipes can never buckle and that the embedded pipe is only a liner inside the concrete structure and has no structural function. This is, in fact, not the case. There has been quite a number of buckling incidents of concrete-embedded pipes.

The buckling event in concrete-embedded pipe may occur in any of the following two stages:

Stage (1): During concreting and in contact with fresh concrete
At this stage, the effects causing buckling of the plastic pipes in fresh concrete environment are:

(1) High external hydrostatic pressure due to high density of concrete compared with water
(2) External hydrodynamic pressure due to concreting and compacting operations
(3) Generation of hydration heat in the concrete and its softening effect on thermoplastic pipe
(4) Possible pre-deformations during transport, storage, and installation

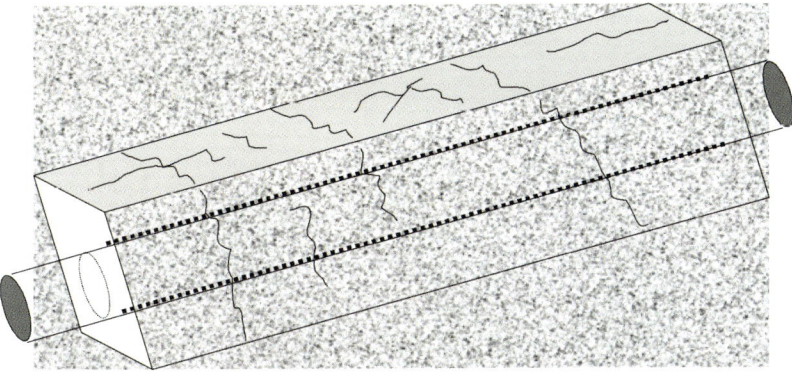

Fig. 4.9. Cracks in the concrete surrounding a plastic pipe.

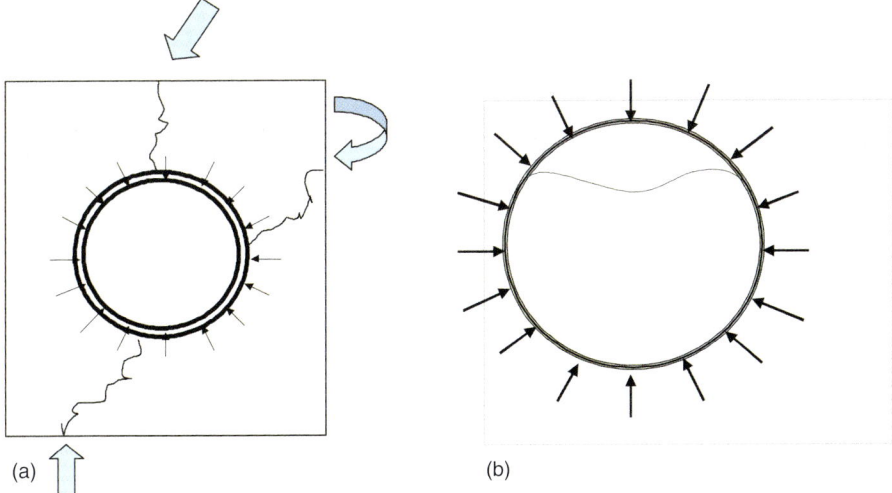

(a)   (b)

Fig. 4.10. (a) Built-up of external hydrostatic pressure in the space between the concrete surrounding a plastic pipe and the pipe due to infiltration of water. (b) Buckling of the concrete-embedded pipe due to external interstitial hydrostatic pressure between the pipe and the surrounding concrete.

Stage (2): In hardened concrete and in service
At this stage, the possible events, effects, and causes of buckling are:

(1) Pre-deformation of the pipe during installation
(2) Cracking of the concrete around the pipe due to shrinkage, settlement, thermal shortening
(3) Diffusion of water from the cracks in the space between the pipe and the concrete
(4) Build-up of external hydrostatic pressure
(5) Buckling of the pipe or large creep deformation

Fig. 4.9 shows the mechanisms involved in the potential buckling of plastic pipes during concreting operations. Fig. 4.10 shows the mechanism of external hydrostatic pressure between the hardened concrete and the embedded pipe in service condition and the resulting buckling event.

4.9   Buckling of the pipe liners

In many cases the existing piping systems should be rehabilitated to meet the functional requirements or to increase their service life. One of the standard methods of rehabilitation is placement of a pipe liner inside the existing piping system. Pipe liners may consist of cured-in-place layers of resin, sprayed resins, or may be in the form of new pipes installed into the old pipe. Thermoplastic liners and liners made of fiber-reinforced plastics are quite common as pipe liners.

Thin liners are normally attached to the existing pipe wall. They function as diffusion and leakage barriers; they may also have structural functions to increase the strength or the stiffness of the system. In the latter function, they should interact with the host pipe system and carry the applied loads in a combined fashion. In some cases, pipe liners are subjected to external hydrostatic pressure. The pressure may be applied to the combined pipe system or may be build up in the space between the liner and the host pipe. The mechanism of pressure built-up discussed in Section 4.8 on concrete embedded pipes can also act in the pipe liners. As the result of this, part of the liners may be detached from the wall of the host pipe and buckle inward. Many cases of inward buckling or pipe liners have been reported in the literature. Buckling of the pipe liners is not of the ovalization type, but in the form of delamination of the pipe liner from the host pipe wall.

4.10   Local buckling due to concentrated loads

In some cases, concentrated forces may act on the pipe system. Examples of such actions are crossing of pipelines over each other, pieces of the embedment stone pieces in contact with the pipe, and pieces of installation woods in the trench, which are sometimes forgotten to be removed. The soil pressure from the side and backfill and the traffic loads push the pipe against the local pieces in contact with the pipe and cause relatively high reaction forces that may produce high localized stresses and strains as well as localized buckling. Pipe on point supports may also experience localized buckling. Fig. 4.11 shows schematically the local buckling of a pipe due to a piece of stone or wood under the pipe.

4.11   Types of buckling and excessive deformation

It is not always easy to distinguish between the buckling instability and the excessive deformation. Buckling is normally referred to a mode of instability of relatively perfect bodies. If the so-called pre-buckling state is not bending-free, then a nonlinear coupling of pre-buckling and post-buckling takes place. In that case a nonlinear interaction is realized.

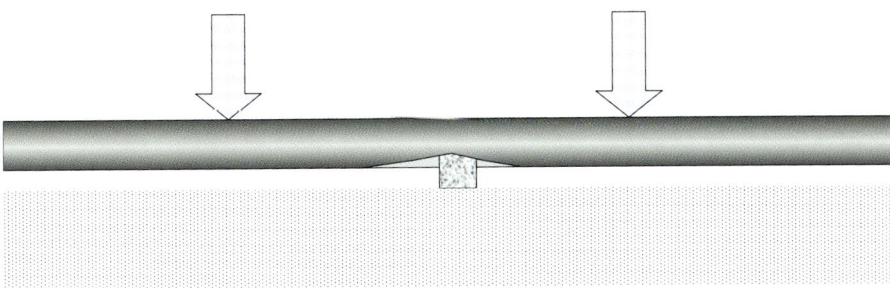

Fig. 4.11. Local buckling of pipe due to point load.

Table 4.1 Types of buckling instability in pipes.

	Buckling								Other
	Axial				Transverse/ring				
Number	Single		Multiple		Single		Multiple		
Extent	Local	Global	Local	Global	Local	Global	Local	Global	
Configuration	Lateral, asymmetric; Axi-symmetric	Lateral/asymmetric; Axi-symmetric	Lateral, asymmetric; Axi-symmetric	Lateral/asymmetric; Axi-symmetric	Dent; Delamination	Flattening	Flattening	Dent; Buckle waves; Flattening; Delamination	Fiber buckling; Face wrinkling; Elephant foot bulging; Total collapse; Dynamic buckling (Flutter)

From the point of view of failure investigation, different notions are used for the buckling and the excessive deformations. Buckling is normally considered as reduction or loss of stiffness and load-carrying capacity of the pipe. Excessive deformation, on the other hand, is a limit on the function of the pipe. Hence buckling may be considered a *strength and stiffness criteria*, while the excessive deformation is related to the *serviceability criteria*. Accordingly, in the standards related to plastic pipes individual safety factors and limiting values are defined for these two categories. Hence, for the purpose of failure investigations, a separate treatment of buckling and excessive deformations proves to be quite useful. Tables 4.1 and 4.2 summarize the various types of buckling instabilities and excessive deformations in plastic pipes.

Table 4.2 Types of excessive deformation in pipes.

	Excessive deformation																								
	Axial								Ring																
Number	Single				Multiple				Single								Multiple								
Extent	Local		Global		Local		Global		Local				Global				Local				Global				
Configuration and orientation	Longitudinal	Transverse	Longitudinal	Bending	Longitudinal	Transverse	Longitudinal	Bending	Swell	Dent	Delamination	Flattening	Swell	Dent	Delamination	Flattening	Swell	Dent	Delamination	Flattening	Swell	Dent	Delamination	Flattening	

4.12   Assessment of buckled pipes

This section contains a basic set of guidelines for failure investigation of plastic pipes as related to the stability assessment. Table 4.3 composed in a compact matrix form, relates the types of instability with the possible causes including mechanical, thermal, chemical, biological, and time factors. In this table, the rows of the matrix designate various possible environmental effects and the columns depict the types of instability, which may occur in the pipe. The probability of an environmental factor being responsible for a specific buckling event is designated by the "+" sign in the matrix. The empty elements of the table are indications that no general correlation may exist between the environmental causes and the buckling events. However, in a specific case, certain correlation (or lack of correlation) may be assesses.

It should be emphasized that the entries in table as well as other assessment tables in this book are based on the knowledge of the author and the inputs from some other experts and the experience gained from various failure investigations. These assessments should only be regarded as general guidelines on which the individual assessments with case-specific modifications and judgment can be based. These tables can be used for failure assessments; they also can be used as a *knowledge base* for creation of an *expert system* for failure diagnosis of plastic pipes.

4.13   Assessment of excessive deformations

Table 4.4 relates the stability types with the possible causes including mechanical, thermal, chemical, biological, and time factors. The table is also presented in a compact matrix form. The probability of an environmental factor being responsible for excessive deformation in plastic pipe is designated by the "+" sign in the matrix. The empty elements of the table are indications that no general correlation may exist between the environmental causes and the buckling events. However, in a specific case, certain correlation (or lack of correlation) may be established.

Table 4.3 Assessment of buckling of plastic pipes (non-exhaustive).

	Buckling																				
	Axial								Transverse/ring								Other				
Number	Single				Multiple				Single			Multiple									
Extent	Local		Global		Local		Global		Local		Global	Local			Global						
Nature and Configuration	Lateral, asymmetric	Axi-symmetric	Lateral, asymmetric	Axi-symmetric	Lateral, asymmetric	Axi-symmetric	Lateral, asymmetric	Axi-symmetric	Dent	Delamination	Flattening	Dent	Buckle waves	Flattening	Flattening	Delamination	Fiber buckling	Face wrinkling	Elephant foot bulging	Total collapse	Dynamic buckling (Flutter)
Influence																					
Material	+	+	+	+	+	+	+	+	+	+	+	+	+	+	+	+	+	+	+	+	+
Mechanical	+	+	+	+	+	+	+	+	+	+	+	+	+	+	+	+	+	+	+	+	+
Internal pressure																					
External pressure	+	+	+	+	+	+	+	+	+	+	+	+	+	+	+	+	+	+	+	+	+
Axial tension																					
Axial compression	+	+	+	+	+	+	+	+		+	+	+	+		+	+	+	+	+	+	+
Bending																					

(continued)

Table 4.3 (*Continued*)

	Buckling																				
	Axial								Transverse/ring								Other				
Number	Single				Multiple				Single			Multiple									
Extent	Local		Global		Local		Global		Local		Global	Local			Global						
Nature and Configuration	Lateral, asymmetric	Axi-symmetric	Lateral, asymmetric	Axi-symmetric	Lateral, asymmetric	Axi-symmetric	Lateral, asymmetric	Axi-symmetric	Dent	Delamination	Flattening	Flattening	Dent	Buckle waves	Flattening	Delamination	Fiber buckling	Face wrinkling	Elephant foot bulging	Total collapse	Dynamic buckling (flutter)
Traffic load											+	+		+	+						
Settlement											+	+		+	+						
Uplift											+	+		+	+						
Production										+		+	+	+		+	+	+			
Impact		+		+		+		+	+	+			+			+	+	+	+	+	+
Vibration																					+
Fatigue											+					+					
Residual stresses											+	+		+	+	+					
Other	+						+														

Thermal	+		+		+	+	+	+		+	+	+		+	+	+	+	+	+	+	
High temperature inside	+		+		+	+	+	+		+	+	+		+	+	+	+	+	+	+	
High temperature outside	+		+		+	+	+	+		+	+	+		+	+	+	+	+	+	+	
UV radiation																					
Fire	+		+		+	+	+	+	+	+	+	+	+	+	+	+	+	+	+	+	
Frost																					
Other																					
Chemical																					
Water																					
Oxygen																					
Acids																					
Alkalis																					
Solvents																					
Oil																					
Benzene																					
Other																					
Service conditions																					
Abrasion																					
Interventions	+	+	+	+	+	+	+	+	+	+	+	+	+	+	+	+	+	+	+	+	+
Other																					
Biological																					
Microbes																					
Animals									+			+									
Other																					
Ageing factors																					
Long-term effects										+						+					

Table 4.4 Assessment of excessive deformations in plastic pipes.

Excessive deformation	Axial								Ring															
Number	Single				Multiple				Single								Multiple							
Extent	Local		Global		Local		Global		Local				Global				Local				Global			
Size and direction	Longitudinal	Transverse	Longitudinal	Bending	Longitudinal	Transverse	Longitudinal	Bending	Swell	Dent	Delamination	Flattening	Swell	Dent	Delamination	Flattening	Swell	Dent	Delamination	Flattening	Swell	Dent	Delamination	Flattening
Influence																								
Material	+	+	+	+	+	+	+	+	+	+	+	+	+	+	+	+	+	+	+	+	+	+	+	+
Mechanical	+	+	+	+	+	+	+	+	+	+	+	+	+	+	+	+	+	+	+	+	+	+	+	+
Internal pressure	+	+	+	+	+	+	+	+	+	+	+	+	+	+	+	+	+	+	+	+	+	+	+	+
External pressure	+	+	+	+	+	+	+	+	+	+	+	+	+	+	+	+	+	+	+	+	+	+	+	+
Axial tension	+	+	+	+	+	+	+	+	+	+	+	+	+	+	+	+	+	+	+	+	+	+	+	+
Axial compression	+	+	+	+	+	+	+	+	+	+	+	+	+	+	+	+	+	+	+	+	+	+	+	+
Bending	+	+	+	+	+	+	+	+	+	+	+	+	+	+	+	+	+	+	+	+	+	+	+	+
Traffic load	+	+	+	+	+	+	+	+	+	+	+	+	+	+	+	+	+	+	+	+	+	+	+	+
Settlement	+	+	+	+	+	+	+	+	+	+	+	+	+	+	+	+	+	+	+	+	+	+	+	+
Uplift	+	+	+	+	+	+	+	+	+	+	+	+	+	+	+	+	+	+	+	+	+	+	+	+
Production																								
Impact																								

Vibration	+	+	+	+	+	+	+	+	+	+	+	+	+	+	+	+	+	+	+
Fatigue	+	+	+	+	+	+	+	+	+	+	+	+	+	+	+	+	+	+	+
Residual stresses	+	+	+	+	+	+	+	+	+	+	+	+	+	+	+	+	+	+	+
Other																			
Thermal																			
High temperature inside	+	+	+	+	+	+	+	+	+	+	+	+	+	+	+	+	+	+	+
High temperature outside	+	+	+	+	+	+	+	+	+	+	+	+	+	+	+	+	+	+	+
UV radiation																			
Fire	+	+	+	+	+	+	+	+	+	+	+	+	+	+	+	+	+	+	+
Frost	+	+	+	+	+	+	+	+	+	+	+	+	+	+	+	+	+	+	+
Other																			
Chemical																			
Water																			
Oxygen																			
Acids																			
Alkalis																			
Solvents																			
Oil																			
Benzene																			
Other																			
Service conditions																			
Abrasion																			
Interventions	+ +	+ +	+ +	+ +	+ +	+ +	+ +	+ +	+ +	+ +	+ +	+ +	+ +	+ +	+ +	+ +	+ +	+ +	+ +
Other																			
Biological																			
Microbes																			
Animals																			
Other																			
Ageing factors																			
Long-term effects	+	+	+	+	+	+	+	+	+	+	+	+	+	+	+	+	+	+	+

4.14   Buckling failure cases

This section contains a number of failure cases related to buckling of plastic pipes. The actual case studies are presented in a compact unified and tabular form. Each of these cases was originally supplemented by a comprehensive documentation containing the details of the case. However, for brevity of presentation, in the present outline of these failure cases the details are not included. Moreover, the information and evidences leading to the hypotheses and the failure judgments are not elucidated.

It is to be emphasized that the failure cases outlined in this section should be considered as the statistical events, which might have been caused by specific circumstances. Hence, the fact that a certain material or pipe type did fail would by no means reflect the weakness or malfunction related to that material. The circumstances leading to these events might have also caused damages in other pipe materials and products as well. One should bear in mind that the material and the pipe are only two factors among the many that could cause a failure to occur. Hence the cases outlined in this section should be treated in a relativistic and demonstrative perspective, and should be devoid from any value judgment and generalization about the specific materials and products.

Each case is documented by a number of sketches and photographs. The list of cases treated in this section is as follows:

Case B-1: Buckling of concrete-embedded pipes (Table 4.5)
Case B-2: Buckling of two mud-carrying pipes (Table 4.6)
Case B-3: Buckling of PE drainage pipe (Table 4.7)
Case B-4: Drainage PE pipe in a water transmission tunnel (Table 4.8)
Case B-5: Buckling of cable protection pipes in a ventilation tunnel (Table 4.9)
Case B-6: Buckling of multi-layer pipes at higher temperatures (Table 4.10)

Table 4.5 Case study of concrete-embedded PE pipes.

Case B-1	Buckling of concrete embedded pipes
Pipe material and dimensions	High-density PE (PE-HD) Two pipes: DN=450 mm; wall thickness, 14 mm One pipe: DN=110 mm; wall thickness, 3.5 mm
Description of the system	Several sewer water plastics pipes (length of pipes 44 m) embedded in concrete inside a steel pipe (internal diameter of 1805 mm) surrounded by the mortar concrete (0–3 mm aggregate) and metallic installation pieces every 2.25 m. The piping system was located under a river in a city. The concreting around the pipes took place by a rotational concrete pump
Failed part	PE-HD pipes
Observed phenomenon	Plastic buckling of several pipes located in a steel pipe
Failure description	Totally collapsed pipe section
Environmental conditions	External alkaline concrete, inside water
Time to failure	Unknown
Tests performed	Buckling under external hydrostatic pressure
Other investigations	Site investigation
Failure cause(s)	Buckling due to the external pressure head caused by concreting, interaction of pipes with lateral elements, and possibly weak pipes. The combined actions of fresh concrete on polymer pipe are due to the higher density of concrete compared with water, dynamic loading due to compacting operations, and softening of the material due to the hydration heat. These actions together with other influences, such as the one in the present case, and also the material weaknesses, installation problems, and insufficient design could lead to pipe buckling failure.
Suggested corrective actions	Higher stiffness, controlled concreting actions (compacting), better installation
Photo documentation	See the related sketches and photographs (Figs 4.12–4.14)

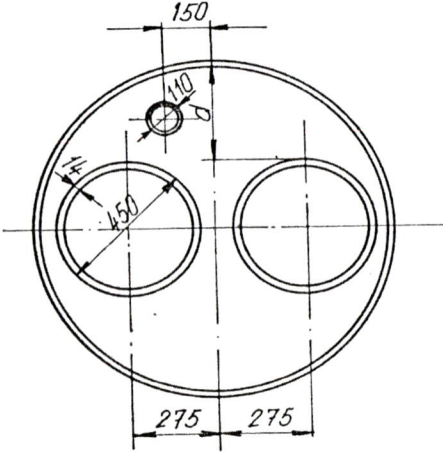

Fig. 4.12. Cross-section showing the outer steel pipe with three thermoplastic pipes inside the steel pipe. The space between the steel pipe and the group of thermoplastic pipes was filled with concrete. Dimensions in the figure are in centimeter.

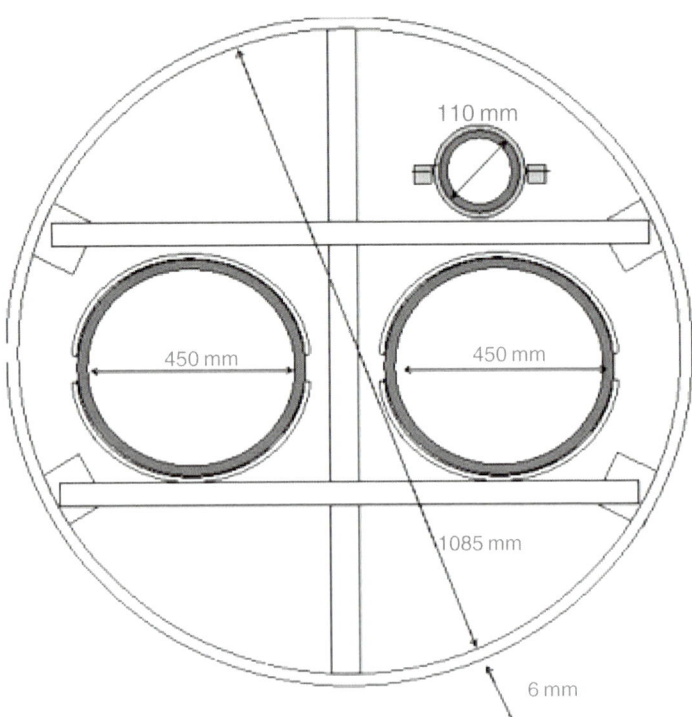

Fig. 4.13. Position of the plastic pipes inside the steel pipe. The figure shows the installation pieces, which were placed at the intervals of about 2250 mm.

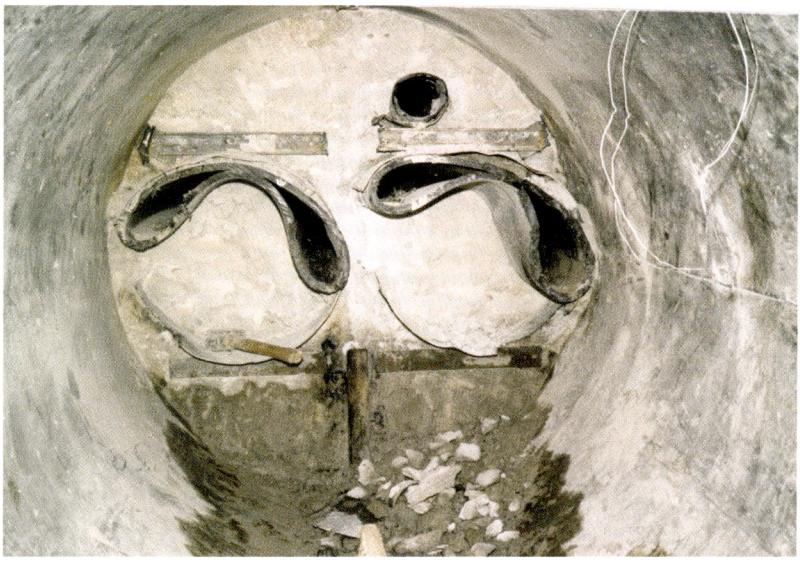

Fig. 4.14. Buckled concrete surrounded thermoplastic pipes inside a steel pipe.

Table 4.6 Case study of buckling of concrete embedded PE pipes.

Case B-2	Buckling of two slurry-carrying pipes
Pipe material and dimensions	High-density PE (PE-HD) pipe (PE80; nominal diameter DN = 160 mm; wall thickness, 6.2 mm)
Description of the system	Two mud-carrying pipes from a container embedded in concrete and locally encased in a 4-cm long ring flange at the transition to the concrete bedding
Failed part	Pipe
Observed phenomenon	Plastic buckling mainly on the crown region
Failure description	Buckled pipe
Environmental conditions	External: concrete and alkaline; internal: slurry
Time to failure	Unknown
Tests performed	Buckling under external hydrostatic pressure
Other investigations	Site investigation
Failure cause(s)	External pressure due to concreting and stiffness jump. The combined actions of fresh concrete on polymer pipe are due to the higher density of concrete compared with water, dynamic loading due to compacting operations, and softening of the material due to the hydration heat. These actions together with other influences, such as the one in the present case, and also the material weaknesses, installation problems, and insufficient design could lead to pipe buckling failure
Suggested corrective actions	Stiffer pipe, evading of complex connections, better installation
Photo documentation	See the related photographs (Figs. 4.15 and 4.16)

Fig. 4.15. Buckled zone of mud-carrying thermoplastic pipes.

Fig. 4.16. Details of buckled pipe.

Table 4.7 Case study of buckling of a drainage PE pipe.

Case B-3	Buckling of PE drainage pipe
Pipe material and dimensions	High-density PE (PE-HD) pipe (nominal diameter DN=280 mm; nominal pressure PN=3.2; pipe wall thickness, 10 mm)
Description of the system	Meteor discharge main water embedded in trench 6-m deep inside a 500-diameter steel pipe and surrounded by broken sand with very small amount of cement (200 kg/m^3)
Failed part	Pipe
Observed phenomenon	Elastic–plastic buckling of the pipe affecting a 6 m of the pipe length
Failure description	Pronounced buckling and total collapse of the pipe section observed with video examination showing a clogged pipe
Environmental conditions	External: confined sand and partly compacted; internal: meteor water
Time to failure	About 2 years
Tests performed	Material tests, buckling under external hydrostatic pressure
Other investigations	Site investigation
Failure cause(s)	The pipe was pushed from the ground level into a steel pipe at a depth of 6 m; the free room between the PE pipe and the steel pipe was filled with sand and was partially compacted. Due to the push-in action in a relatively short trench length the pipe had been ovalized. By filling of the sand and its partial compaction the pipe was subjected to non-uniform external pressure mainly from the lateral side, while the top was probably empty. Due to this action the pipe had buckled
Suggested corrective actions	Better installation of the lining
Photo documentation	See the related sketches and photographs (Figs. 4.17–4.20)

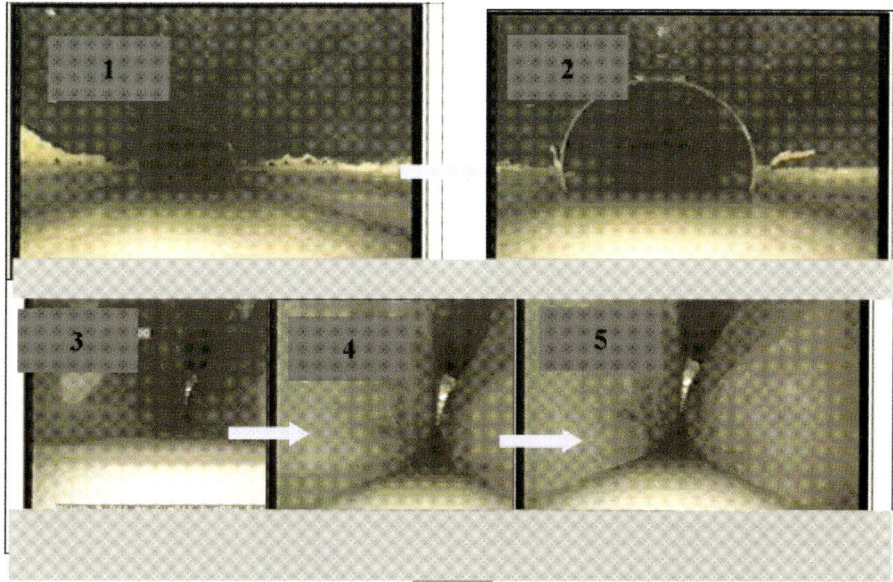

Fig. 4.17. Sequence of video recording of the buckled pipe; the buckling affected a length of several meters.

Fig. 4.18. The excavated buckled pipe. The buckling deformation is partially recovered.

Fig. 4.19. Picture of the excavated pipes after partial buckling recovery; the figure shows the remaining plastic deformation.

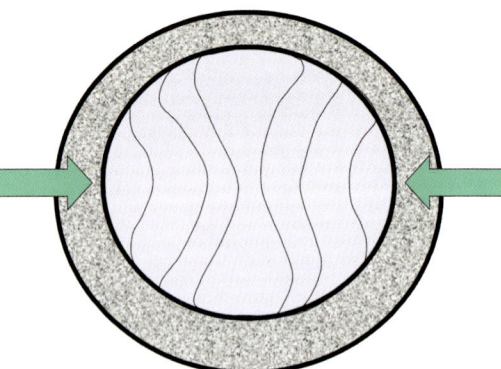

Fig. 4.20. Schematics of the hypothesis about the buckling cause.

Table 4.8 Case of buckling of a PE drainage pipe in a tunnel floor.

Case B-4	Drainage PE pipe in a water transmission tunnel
Pipe material and dimensions	High-density PE (PE-HD) pipe (nominal diameter DN = 600 mm; standard diameter ratio SDR = 32, pipe wall thickness about 20 mm)
Description of the system	Two drainage water pressure pipes positioned in concrete floor of the tunnel; length about 2.2 km. Pipes were embedded in a canal with round gravel around in the bottom part of a drainage tunnel with the inner diameter of 4 m. There were two such pipes. The two pipes were used as dry weather drainage pipes. Next to the pipe in question was a small fresh pressure water pipe. The whole system was surrounded by a concrete floor
Failed part	Pipe
Observed phenomenon	Description of the failure case: Buckling of the pipe along about 15 m. Long and circumferential cracks in the floor; heaving of the floor mainly over the buckled pipe. After boring of the pipe at the buckled place external water jet came into the pipe and lasted several minutes
Failure description	Pipe was buckled inward at various places Longitudinal cracks on the concrete floor, indications of heaving Water came out of the bored hole from outside into the pipe Water pressure could be built up under the tunnel floor Concrete not quite water-tight
Environmental conditions	Not expandable, mixed, water conducting Ground water table: maximum 35 m
Time to failure	Newly installed
Tests performed	None
Other investigations	Site investigation
Failure cause(s)	Weak pipe; action of hydrostatic pressure; pre-deformation (before installation and during installation, concreting). Three factors were considered: (1) external water pressure from the surrounding tunnel environment, (2) rock pressure, (3) heaving of the piping system. The hypothesis set forth by the consulting engineering company was that the buckling of the pipe occurred due the external water pressure and the lack of sufficient buckling strength due to lack of proper dimensioning
Suggested corrective actions	Stronger pipe, better installation condition
Photo documentation	See the related sketches and photographs (Figs. 4.21–4.24)

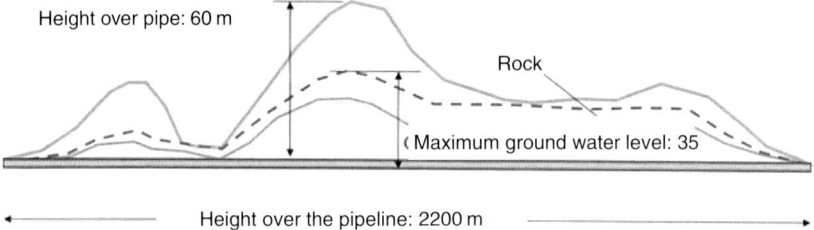

Height over pipe: 60 m

Rock

(Maximum ground water level: 35

Height over the pipeline: 2200 m

Fig. 4.21. Longitudinal profile of the tunnel and the piping system.

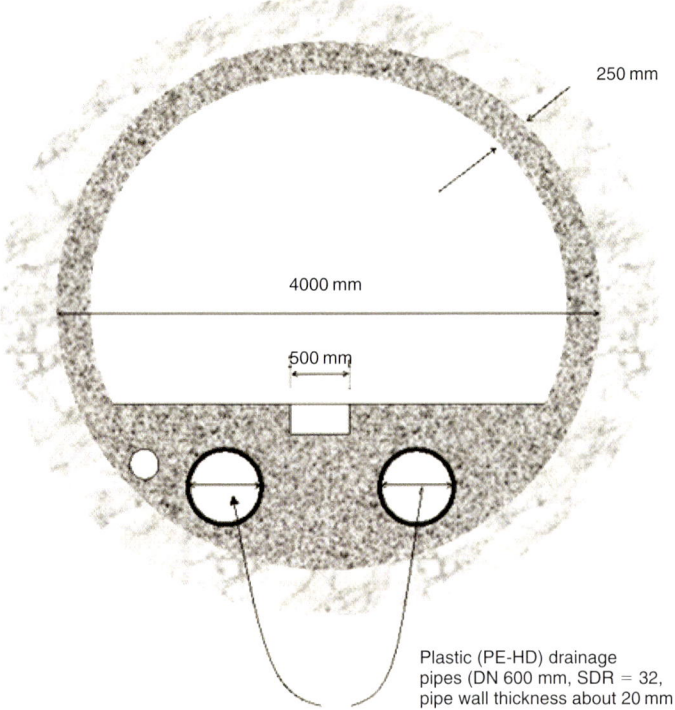

250 mm

4000 mm

500 mm

Plastic (PE-HD) drainage
pipes (DN 600 mm, SDR = 32,
pipe wall thickness about 20 mm

Fig. 4.22. Cross-section of the tunnel with concrete-embedded thermoplastic pipes.

Fig. 4.23. Buckled pipe seen through a cut opening. The picture shows the jet of water flowing into the pipe from the surrounding concrete embedment.

Fig. 4.24. Buckled pipe seen through a cut opening.

Table 4.9 Case study of buckling of cable protection pipes.

Case B-5	Buckling of cable protection pipes in a ventilation tunnel
Pipe material and dimensions	Rigid polyvinyl chloride (PVC-H) pipe (nominal diameter DN=84 mm; wall thickness, 2.1 mm)
Description of the system	Four PVC-H cable protection pipes embedded in concrete in the bottom region of a ventilation tunnel with a strong inclination of about 45°. Pipes were placed at a distance of about 150 mm from each other on a 2-mm thick water sealing membrane and a 20-mm thick drainage matt
Failed part	Pipe
Observed phenomenon	Buckling of the pipes during concreting operations
Failure description	Buckled pipe
Environmental conditions	External: concrete bedding
Time to failure	During installation
Tests performed	Buckling under external hydrostatic pressure
Other investigations	Calculated critical concrete height during concreting: about 3.6 m
Failure cause(s)	External pressure due to concreting
Suggested corrective actions	Stronger pipe, better installation conditions
Photo documentation	See Fig. 4.25

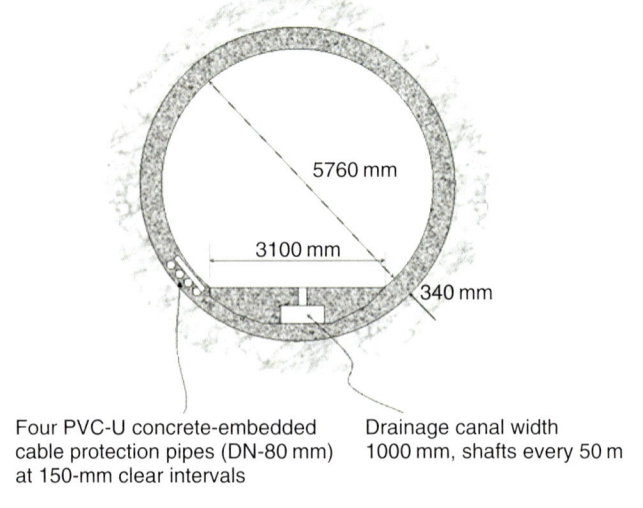

Four PVC-U concrete-embedded
cable protection pipes (DN-80 mm)
(a)   at 150-mm clear intervals

Drainage canal width
1000 mm, shafts every 50 m

Fig. 4.25. (a) Cross-section of the tunnel with four cable protection pipes at the bottom region.

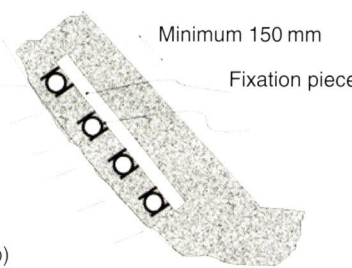

Minimum 150 mm

Fixation piece

(b)

Fig. 4.25. Contd. (b) Magnified detailed view of the zone of four cable protection pipes.

Table 4.10 Typical case of buckling of multi-layer pipes.

Case B-6	Buckling of multi-layer pipes at higher temperatures
Pipe material and dimensions	A multi-layer pipe may generally consist of two or more layers. The layers may all consist of polymeric materials or a layer may be metallic. The function of metallic layer is to function as a diffusion barrier. The metallic layer can also contribute to a higher strength and stiffness. The layer may be attached to each other by thin layers of glue materials
Description of the system	Multi-layer piping systems have applications in hot water transport systems
Failed part	Pipe
Observed phenomenon	Possible inward buckling of the inner layer at higher temperatures. Possible kinking of the pipe as a whole
Failure description	Delamination of the internal layer and its inward buckling
Environmental conditions	Pressure hot water inside; the outside environment dependent on the installation condition
Time to failure	During installation or in service
Tests performed	This case was typical of inward buckling mode of multi-layer pipes. For each particular failure case appropriate investigation program should be carried out.
Other investigations	See above comment.
Failure cause(s)	Difference in thermal coefficient of expansion of the layers, weak bonding between the layers, insufficient design
Suggested corrective actions	Design improvement, strength of bonding, better installation
Photo documentation	See Figs. 4.26 and 4.27

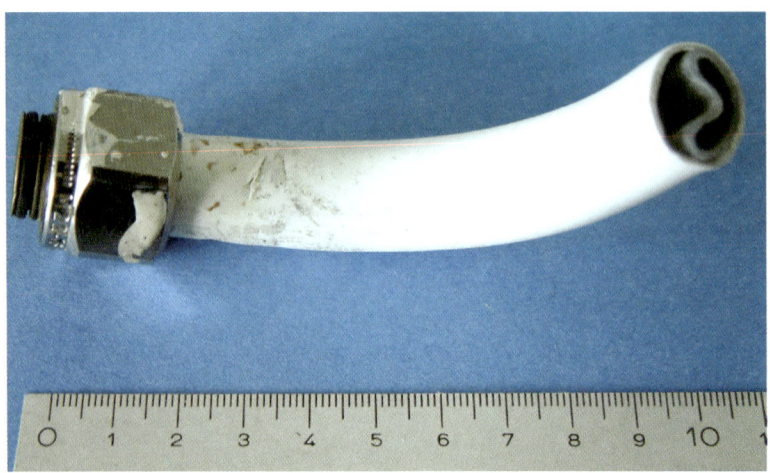

Fig. 4.26. A delamination buckling of a bent multi-layer pipe.

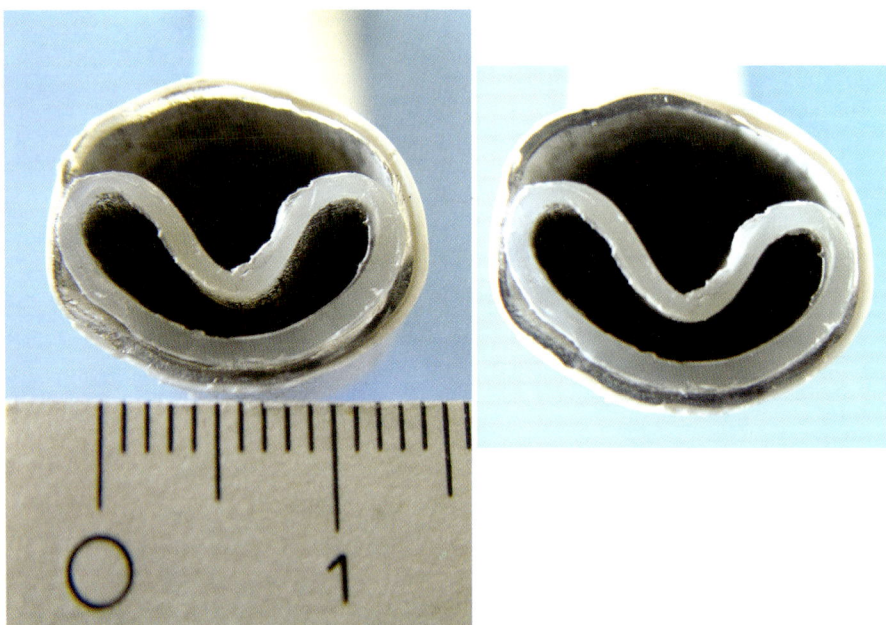

Fig. 4.27. Sectional picture of a multi-layer pipe; the inward buckling of the inner layer.

4.15 Preventive measures against buckling and excessive deformation

Buckling and excessive deformation of pipelines is a potential mode of failure, which may mainly occur in sewer pipes, in cable protection pipes, in drainage pipes, and in pipes under elevated temperature or axial compression. Pressure pipe systems may also experience buckling failure in the empty condition. Buckling and excessive deformation of pipes may lead to malfunction of the pipe system and hence are to be avoided. In sewer and drainage systems, the flow of fluids inside the pipe can be partially or totally stopped. In cable protection pipes, buckling and lateral deformation would prevent the placement of the cable inside of the pipe. In pipe liners, buckling occurs as delamination of the liner from the pipe wall that might reduce the fluid flow or even clog the pipe. Buckling of pipes can not only jeopardize the function of the system, but may also have other costly consequences. For example, in the concrete embedded drainage or cable protection pipes in tunnels, the negative consequences of buckling and excessive deformations are almost imaginable.

Plastic pipes, in comparison with pipes made of other materials, are generally more flexible. In addition to that, the stiffness of plastic pipes can be strongly influenced by temperature, chemicals, and time factor. Creep buckling of plastic pipes is the result of reduced stiffness under sustained load and thermal as well as chemical effects in the course of time. Initial deformations and geometric imperfections may reduce the buckling stability and increase the probability of creep-like excessive deformations.

The buckling failure of plastic pipes may be traced back mainly to the following factors:

(1) Design and planning errors (insufficient dimensioning, lack of sufficient safety factors)
(2) Material and pipe weakness (material weakness, weak composite construction, weak pipe section)
(3) Initial deformations and geometric imperfections
(4) Installation errors (improper bedding conditions, concentrated forces)
(5) Fixed points and change of directions
(6) Unpredicted loads
(7) Inappropriate service conditions
(8) Improper repairs
(9) External interventions

The measures against buckling and excessive of plastic pipes should assure that the above-mentioned sources of error are eliminated or taken into consideration. The bodies responsible for a pipe system should follow the whole service life of the pipe from the planning stage to its repair and replacement. For this purpose, appropriate quality control and health monitoring systems should be used. Furthermore, it is important to take into consideration the various load cases, applicability of the material, and the limits of the pipe service life. The best approach in arriving at preventive measures against potential is to plan the pipe system for the conceived applications, to choose the right material, to monitor the pipe system, and to take the corrective measures during the service life of the pipeline.

5
Weathering, color, and dimensional changes

5.1 Weathering

Weathering is a universal phenomenon occurring in plastic materials subjected to external environment. The weathering effects may be caused by the exposure of the material to the outdoor thermal, radiation, and chemical conditions. Weathering may manifest itself in various forms; it can cause change of color, change of texture, change of dimensions, weakening of molecular chains, change of mechanical, electrical, and physical properties. Weathering may lead to complete deterioration of the material. Weathering factors include: heat, light, oxygen, water/humidity, and pollutants. The duration of action of each of these effects may range from short-term to long-term periods.

The most important changes in materials properties include mass loss, change of mechanical, thermal, electrical, and optical properties, and changes of molecular weight, chemical composition, color, and surface properties of the material. Table 5.1 summarizes the effects of various agents on the weathering of polymers and polymer products including plastic pipes.

Polymeric materials and plastic pipes exposed to direct sunlight containing ultraviolet (UV) rays may undergo weathering and, in some extreme situations, total degradation. The UV energy absorbed by plastics can excite photons, which then create free radicals. In the presence of oxygen the free radicals form oxygen hydro peroxides that can break the double bonds of the backbone chain leading to a brittle structure. The presence of impurities in plastics acts as receptors of UV rays, thus causing chemical reactions and oxidation of the plastics. This process is often called *photo-oxidation*. In the absence of oxygen degradation due to the cross-linking process can also take place.

All types of UV rays can cause a photochemical effect within the polymer structure, which can lead to degradation of the material. The main parameters related to the effect of UV rays on polymers are the stratospheric ozone, clouds, altitude, the position of the sun height, and reflection. Moreover, the ambient temperature and humidity would accelerate the intensity of the UV actions. Other factors other than light such as biological agents may also cause weathering in polymers.

The main visible effects of weathering degradation are color shift on the surface of the material. Moreover, due to weathering, physical, and mechanical properties of polymers undergo changes. One of the major changes is the loss of ductility in the

Table 5.1 The effects of various agents on the weathering of polymers and polymer products including plastic pipes.

Effect	Resulting actions
Light	UV part of solar radiation (wavelength below 400 nm) initiates photochemical reactions leading to degradation of polymer. The absorbed UV radiation causes dissociation of chemical bonds (mostly C−C and C−H) in the polymer molecule. The resulting chemical changes are chain scission (cleavage into smaller molecules), cross-linking (elimination of smaller molecules), formation of double bonds in the main chain, and hydrolysis
Temperature (heat)	Higher temperatures cause reaction kinetics, which leads to polymer degradation
Oxygen	Oxygen together with UV radiation causes photo-degradation or photo-oxidation
Water and humidity	Polymers absorb water and swell. Upon drying the surface layer contacts but its contraction is hindered by the underlying layers, which are still swollen. This results in tensile stresses on the surface of the material, which may lead to cracking. During water absorption, compressive stresses dominate the external part while the internal part is in tensile stress. During water resorbtion, the situation is reverse. Dimension change and contraction of polymers and the resulting stresses is aggravated by solar radiation. The photochemical reactions caused by solar radiation cause embrittlement of the material surface; hence tendency to surface cracking due to tensile stresses during resorbtion is further promoted
Pollutants	The pollutants causing polymer deterioration include inert particulate matter, biologically active materials, and reactive gases. Air movement also plays a major role in the action of pollutants

UV: ultraviolet.

weathered components and the pronounced embrittlement. The above effects are predominantly on the surface layer of the material. However, stress concentrations and the stiffness gradients caused by the highly brittle nature of some commodity plastics may well lead to a complete failure of the component.

There are several ways of avoiding UV degradation in plastics; these include the use of the so-called stabilizers, UV absorbers, and UV blockers. The stabilizers prevent the chemical reaction of the radicals. It is also possible to add antioxidants to some plastics to avoid photo-oxidation. For many outdoor applications, addition of a low percent of carbon black will provide the protection for the structure by the blocking process. Other pigments such as titanium dioxide can also be effective. In the extruded plastic pipes, UV stabilizers and blockers are added as carbon black and pigments to protect them against the effect of UV rays.

Table 5.2 Types of color changes in plastic pipes.

	Color change											
	Through thickness		**External**		**Internal**		**Extent**		**Sedimentation**		**Persistence**	
Specification	Whole section	Part section	Whole section	Part section	Whole section	Part section	Local	Global	Yes	No	Permanent	Removable
Extent	Whole perimeter	Part of section	Whole perimeter	Part of the surface	Whole perimeter	Part of the surface	Local	Global	Yes	No	Permanent	Removable

5.2   Color changes

Change of color of plastic pipes subjected to chemicals and UV rays one of the salient manifestations of physical and chemical changes in material. Colors changes may be caused by chemical interactions or physical phenomenon such as diffusion of other materials into the pipe wall. Color is brought about by thermal ageing of the pipes, UV radiation, and other agents related to the weathering of polymeric materials. Color changes may also be caused by diffusion of substances into the plastic material.

Diffusion is defined as the penetration of the fluid or gas and damp into the material and it's Brownian motion inside the molecular structure. A special type of diffusion is called *self-diffusion*. The speed of diffusion depends on the chemical structure, the morphology of polymer, the temperature, the stress field, and the size of the molecules. Also in the case of fiber-reinforced polymers, the fibers may act as canals for the diffusing substance. Table 5.2 presents the types of color changes, which may occur in plastic pipes. Further aspects of color changes are discussed in Chapter 7 dealing with the corrosion of plastic pipes.

5.3   Dimension changes

Dimension change of plastic materials and pipes is defined as macroscopic increase in the volume of a polymeric material as the result of diffusion of low molecules (especially solvents, water) into the surface of the body. Dimensional changes occurring in plastic pipes consist mainly of swelling of the pipe due to infiltration of an external

Table 5.3 Types of dimensional changes in plastic pipes.

	Dimensional change										
Specification	Pipe				Fitting				Section		Persistence
Extent	Symmetric	Non-symmetric	Axial	Axial and lateral	Symmetric	Non-symmetric	Axial	Axial and lateral	Whole thickness	Partial	Permanent · Recoverable

fluid into the pipe. The index of dimension change is the change of the material density or the dimensions of the body. Due to dimension change the molecular bond of the polymer will be changed. In the case of penetration of solvents into the molecular structure, the molecular bond will be partly or as the whole reduced. Related to dimensional changes, the following rules are valid:

(1) Dimension change is more pronounced in the linear molecules.
(2) The dimension change decreases with increase of cross chains.
(3) Dimension change in an anisotropic material occurs anisotropically.
(4) Low molecules are easier dissolved as the high molecules.
(5) Amorphous polymers dissolve easier than the crystalline polymers.
(6) Through dimension change, physical, and mechanical properties of material change.

Table 5.3 summarizes the types of dimensional changes in plastic pipes. Further aspects of dimension changes are discussed in Chapter 7 dealing with the corrosion of plastic pipes.

5.4 Assessment of color and dimensional changes

This section contains a basic set of guidelines for failure investigation of plastic pipes as related to the stability assessment. Tables 5.4 and 5.5, in respective fashion, relate various types of color and dimensional changes in plastic pipes with the possible causes including mechanical, thermal, chemical, biological, and time factors. These tables are presented in a compact matrix form. The rows of each matrix designate various possible environmental effects and the columns depict the types of color and dimensional changes, which may occur in the pipe. The probability of an environmental factor being responsible for a specific buckling event is designated by the "+" sign in the matrix. The empty elements of the tables are indications that no general correlation may exist between the environmental causes and the failure events. However, in a specific case, certain correlation (or lack of correlation) may be established.

It should be emphasized that the entries in Tables 5.4 and 5.5 as well as other assessment tables in this book, are based on the knowledge of the author and the inputs from

Table 5.4 Assessment of color change in plastic pipes (non-exhaustive).

	Color change											
Specification	Through thickness		External		Internal		Extent		Deposition?		Persistence	
Sectional position	Whole section	Part section	Whole section	Part section	Whole section	Part section	Local	Global	Yes	No	Permanent	Removable
Extent	Whole perimeter	Part of section	Whole perimeter	Part of the surface	Whole perimeter	Part of the surface	Local	Global	Yes	No	Permanent	Removable
Influence												
Material	+	+	+	+	+	+	+	+	+	+	+	+
Mechanical												
Internal pressure												
External pressure												
Axial tension												
Axial compression												
Bending												
Traffic load												
Settlement												
Uplift												
Production	+	+	+	+	+	+	+	+	+	+	+	+
Impact												
Vibration												
Fatigue												
Residual stresses												
Other	+	+	+	+	+	+	+	+	+	+	+	+
Thermal												
High temperature inside	+		+	+	+	+	+	+	+	+	+	+
High temperature outside	+	+	+			+	+	+	+	+	+	+
UV radiation			+	+			+	+	+	+	+	+

(*Continued*)

Table 5.4 (*Continued*)

Color change

Specification	Through thickness		External		Internal		Extent		Deposition?		Persistence	
Sectional position	Whole section	Part section	Whole section	Part section	Whole section	Part section	Local	Global	Yes	No	Permanent	Removable
Extent	Whole perimeter	Part of section	Whole perimeter	Part of the surface	Whole perimeter	Part of the surface	Local	Global	Yes	No	Permanent	Removable
Fire												
Frost	+	+	+	+			+	+	+	+	+	+
Other	+	+	+	+	+	+	+	+	+	+	+	+
Chemical												
Water												
Oxygen	+	+	+	+	+	+	+	+	+	+	+	+
Acids	+	+	+	+	+	+	+	+	+	+	+	+
Alkalis	+	+	+	+	+	+	+	+	+	+	+	+
Solvents	+	+	+	+	+	+	+	+	+	+	+	+
Oil	+	+	+	+	+	+	+	+	+	+	+	+
Benzene	+	+	+	+	+	+	+	+	+	+	+	+
Other	+	+	+	+	+	+	+	+	+	+	+	+
Service conditions												
Abrasion												
Interventions	+	+	+	+	+	+	+	+	+	+	+	+
Other	+	+	+	+	+	+	+	+	+	+	+	+
Biological												
Microbes	+	+	+	+	+	+	+	+	+	+	+	+
Animals												
Other												
Ageing factors												
Long-term effects	+	+	+	+	+	+	+	+	+	+	+	+

Table 5.5 Assessment of dimensional changes in plastic pipes (non-exhaustive).

Specification	Pipe				Fitting				Section		Persistence	
Extent	Symmetric	Non-symmetric	Axial	Axial and lateral	Symmetric	Non-symmetric	Axial	Axial and lateral	Whole thickness	Partial	Permanent	Recoverable
Influence												
Material	+	+	+	+	+	+	+	+	+	+	+	+
Mechanical												
Internal pressure	+		+	+	+		+	+	+	+	+	
External pressure												
Axial tension												
Axial compression												
Bending												
Traffic load												
Settlement												
Uplift												
Production												
Impact												
Vibration												
Fatigue												
Residual stresses												
Other												
Thermal												
High temperature inside		+	+	+		+	+	+	+	+	+	
High temperature outside												
UV radiation												
Fire		+	+	+		+	+	+	+	+	+	
Frost												
Other												
Chemical												
Water	+	+	+	+	+	+	+	+	+	+	+	+
Oxygen		+	+	+	+	+	+	+	+	+	+	
Acids		+	+	+	+	+	+	+	+	+	+	
Alkalis		+	+	+	+	+	+	+	+	+	+	
Solvents		+	+	+	+	+	+	+	+	+	+	
Oil	+	+	+	+	+	+	+	+	+	+	+	

(*Continued*)

Table 5.5 (*Continued*)

	Dimensional change											
Specification	Pipe				Fitting				Section		Persistence	
Extent	Symmetric	Non-symmetric	Axial	Axial and lateral	Symmetric	Non-symmetric	Axial	Axial and lateral	Whole thickness	Partial	Permanent	Recoverable
Benzene	+	+	+	+		+	+	+	+	+	+	
Other	+	+	+	+	+	+	+	+	+	+	+	+
Service conditions												
Abrasion												
Interventions												
Other												
Biological												
Microbes	+	+	+			+	+	+	+	+	+	
Animals												
Other												
Ageing factors												
Long-term effects	+	+	+			+	+	+	+	+	+	

some other experts and the experience gained from various failure investigations. These assessments should be regarded only as general guidelines on which the individual assessments with case-specific modifications and judgment may be based. These tables can be used for failure assessments; they also can be used as a *knowledge base* for creation of an *expert system* for failure diagnosis of plastic pipes.

5.5 A case study related to color and dimensional changes

In this section a failure case study related to color and dimension changes in a thermoplastic pipe system is treated. The pipe system consisted of two industrial sewer polypropylene pipes, which were placed in a utility tunnel for the discharge of chemical wastes in a chemical installation. The piping system had shown some indication of lack of tightness. The site visit revealed color and dimensional changes in many positions along the piping system. As part of investigations, internal pressure tests on the pipe samples at elevated temperature was carried out. The internal pressure test of a pipe samples related to this case is dealt with in Chapter 3.

Table 5.6 Case of an industrial pressure pipe system made of polypropylene (see also the same case in Chapter 3 on fracture of pipes).

Failure case WCD-1	Sewerage industrial waste pressure thermoplastic pipeline
Pipe material and dimensions	Polypropylene (PP-C) pipe (nominal diameter DN = 400 mm, nominal wall thickness = 15.4 mm, nominal pressure PN = 4 bar)
Description of the system	Two-polypropylene exposed pipes supported at intervals of about 3 m inside of a tunnel. The metallic supports were fixed to the concrete wall of the tunnel. The length of the pipeline between the fixed punt was about 300 m. The pipeline was periodically cleaned.
Failed part	The pipe and the butt welds
Observed phenomenon	Color change and internal cracking, change of dimensions, loss of water tightness.
Failure description	Color changes mostly at the crown region and intensified around the butt welds; and a number of longitudinal cracks and strong color darkening around the crack.
Environmental conditions	*External*: Air, the pipeline located in a concrete tunnel *Internal medium*: A mix of chemical agents (T = 15–30 °C)
Time to failure	After 15 years in service
Tests performed	OIT
Other investigations	Site investigation
Failure cause(s)	Action of chemical agents, thermal degradation
Suggested corrective actions	Improvement of resistance to chemicals
Photo documentation	See Figs. 5.1–5.3

OIT: oxygen induction time.

Table 5.6 summarizes the failure case study related to this piping system. Fig. 5.1 shows a photograph of the two parallel pipes and the point supports of the pipes in the installation tunnel. Fig. 5.2 shows the color change at the crown region of pipes at the butt fusion. Finally, Fig. 5.3 shows an inside view of the pipe with the color changes in the crown region.

Fig. 5.1. Two polypropylene pipes on local supports with change of color at the crown region.

Fig. 5.2. The color change at the crown region of pipes at the butt fusion.

Fig. 5.3. Inside of one of the polypropylene pipes showing a change
of color at the crown region.

5.6   Preventive measures against weathering, color, and dimensional changes

Weathering, color, and dimensional changes are directly related to the material type and environmental effects. These phenomena may occur due to exposure of plastic pipe to UV radiation, high temperatures, and chemical agents. Certain types of additives including pigments and stabilizators can help to improve the durability of plastics under the above-mentioned factors. Addition of powder and reinforcing materials can also help to the plastic pipe to maintain its dimensions and to increase the resistance against dimensional changes. Weathering and change of properties may occur in the pipes, which are stored for a relatively long time at the outdoor environment with UV effects and aggressive medium. Hence, to evade or reduce the weathering effects, care should be taken to store the pipes in a protected environment.

6

Voids, blisters, and delaminations

6.1 Categories of local material failure

In this chapter a number of phenomena related to local material failure in plastic and composite pipes are treated. These phenomena may be caused by the processing of the material or may be caused by the handling and service conditions.

Voids are macro and micro-cavities, which may exist in the material or be caused by the production procedure, service conditions, and chemicals. Voids are in fact empty spaces inside the pipe wall. Voids may sometimes be filled with gases other than the air.

Blisters are visible voids near the pipe internal and external surface. In some cases, blisters are in the form of large bulging of some layers of a laminated pipe.

Delamination is debonding of the layers of composite pipes or detachment of a pipe liner from the host pipe.

Table 6.1 summarizes the inherent material imperfections or the caused local damages in plastic pipes.

6.2 Assessment of local damages

This section contains a basic set of guidelines for failure investigation of plastic pipes as related to the local imperfections and damages. Table 6.2 relates various types of local damages in plastic pipes with the possible causes including mechanical, thermal,

Table 6.1 Material imperfections or the induced local damages in plastic pipes.

	Voids/Blisters/Delamination																							
	Voids				External blister								Internal blister								Delamination			
Extent					Local				Global				Local				Global				Local		Global	
Size	Micro	mm size	cm size	Larger	Micro	mm size	cm size	Larger	Micro	mm size	cm size	Larger	Micro	mm size	cm size	Larger	Micro	mm size	cm size	Larger	Inside	Outside	Inside	Outside

Table 6.2 Assessment of local damages in plastic pipes (non-exhaustive).

Voids/Blisters/Delamination

	Voids – Local				External blister – Local				External blister – Global				Internal blister – Local				Internal blister – Global				Delamination – Local		Delamination – Global	
Extent / Size	Micro	mm size	cm size	Larger	Micro	mm size	cm size	Larger	Micro	mm size	cm size	Larger	Micro	mm size	cm size	Larger	Micro	mm size	cm size	Larger	Inside	Outside	Inside	Outside
Influence																								
Material	+	+	+	+	+	+	+	+	+	+	+	+	+	+	+	+	+	+	+	+	+	+	+	+
Mechanical																						+	+	+
Internal pressure																							+	
External pressure																								+
Axial tension																								
Axial compression																						+	+	+
Bending																								
Traffic load																								
Settlement																								
Uplift																								
Production	+	+	+	+	+	+	+	+	+	+	+	+	+	+	+	+	+	+	+	+	+	+	+	+
Impact									+	+	+	+										+		
Vibration																								
Fatigue	+	+	+	+	+	+	+	+	+	+	+	+	+	+	+	+	+	+	+	+	+	+	+	+
Residual stresses																								
Other																								

Thermal
High temperature
  inside
High temperature
  outside
UV radiation
Fire
Frost
Other
Chemical
Water
Oxygen
Acids
Alkalis
Solvents
Oil
Benzene
Other
Service conditions
Abrasion
Interventions
Other
Biological
Microbes
Animals
Other
Ageing factors
Long-term effects

chemical, biological, and time factors. This table is presented in compact matrix form. The rows of the matrix designate various possible environmental effects and the columns depict the types of local damages, which may occur in the pipe. The probability of an environmental factor being responsible for a specific damage event is designated by the "+" sign in the matrix. The empty elements of the tables are indications that no general correlation may exist between the environmental causes and local damages. However, in a specific case, certain correlation (or lack of correlation) may be established.

It should be emphasized that the entries in Table 6.2 are based on the knowledge of the author and the inputs from some other experts and the experience gained from various

Table 6.3 Case of large blisters and delaminations in a GFRP pipe system.

Failure case BD-1	Blisters in concrete-embedded GFRP pipeline for water turbines
Pipe material and dimensions	GFRP (short fiber reinforced), pipe (diameter: 1250 mm, wall thickness: about 11 mm)
Description of the system	Concrete-embedded pressure pipe for water power turbines; the upper part of the piping system with free water level. The upper part of the piping system was pressure less with relatively large inclination with horizontal direction. This part consisted of concrete-embedded GFRP pipes having diameters 1250, 1000, and 800 mm. The slope of this part was about 10%. The piping system was located in a mountainous area on a relatively sharp slope. The concrete embedment consisted of a block with side dimensions of about 2 m × 2 m. The length of the pipeline was 1260 m.
Failed part	Pipe
Observed phenomenon	Large blisters (typical dimensions: width 500 mm, height 50 mm) in the pipe wall thickness filled with milky fluid.
Failure description	Blisters mainly located at the bottom and spring line regions, strong delamination of layers.
Environmental conditions	*External*: concrete with alkaline content; *internal*: mountain water for water turbine.
Time to failure	About 27 years
Tests performed	Tests performed: chemical tests on entrapped fluid, plasma emission spectrometry (IOP-OES), glass composition test, tensile test.
Other investigations	Site investigation: during site investigation, the inner part of the pipe was visited. Several blisters were punctured, upon which the milky water came out in a jet fashion.
Failure cause(s)	Osmosis, ageing
Suggested corrective actions	Choice of a more appropriate resin
Photo documentation	See Figs. 6.1 and 6.2

GFRP: glass-fiber-reinforced polyester.

failure investigations. These assessments should only be regarded as general guide-lines on which the individual assessments with case-specific modifications and judgment may be based. These tables can be used for failure assessments; they also can be used as a *knowledge base* for creation of an *expert system* for failure diagnosis of plastic pipes.

6.3   Investigation of failure cases

In this section two failure cases related to the phenomena voids, blisters, and delaminations are outlined. The first case deals with the failure of a concrete-embedded glass-fiber-reinforced pipe in a water turbine piping system. Some features of this case are outlined in Table 6.3. Figs. 6.1 and 6.2 show pictures of the pipe with large blisters filled with water under internal pressure. The second case relates to the blisters on the surface of a thermoplastic pipe. Table 6.4 summarizes the available information on this case. Figs. 6.3 and 6.4 show the pipe surface with blisters.

Fig. 6.1. One of the several large blisters observed at the inside of the glass-fiber-reinforced polyester pipe.

Fig. 6.2. Water jet springing from one of the blisters upon puncturing the blister.

Table 6.4 Case of blisters on the inside surface of a PE pipe.

Failure case BD-1	Blisters on the surface of a PE pipe
Pipe material and dimensions	PE-HD pipe, wall thickness: 15 mm
Description of the system	PE pipe without inside fluid
Failed part	Pipe
Observed phenomenon	Blisters on the internal surface
Failure description	Blisters and surface irregularities
Environmental conditions	
Time to failure	
Tests performed	IR spectrometry
Other investigations	Visual examinations
Failure cause(s)	In material external substance was found
Suggested corrective actions	Material production improvement
Photo documentation	See Figs. 6.3 and 6.4

PE: polyethylene; PE-HD: polyethylene, high density.

Fig. 6.3. Blisters on the inside surface of the polyethylene pipe.

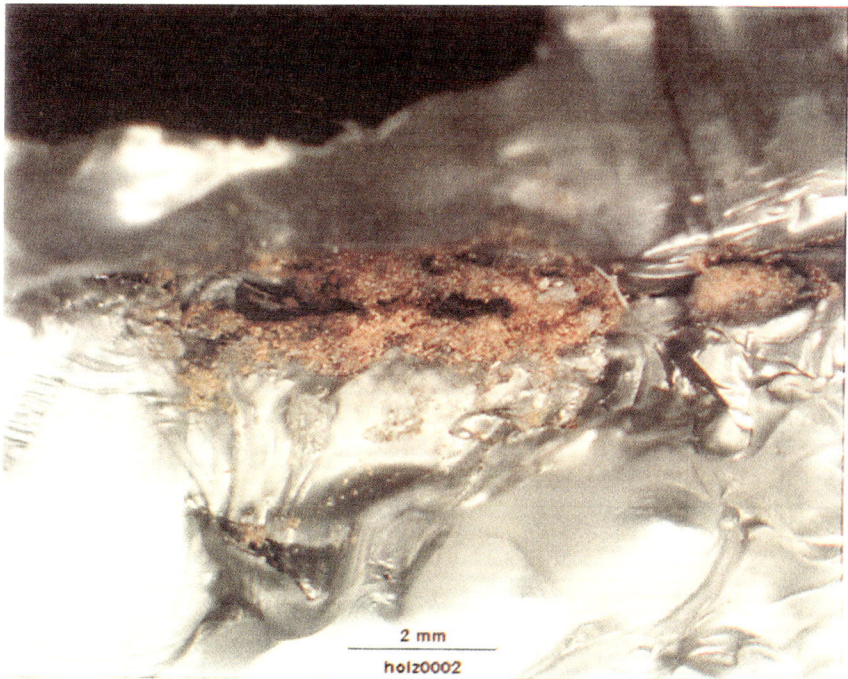

Fig. 6.4. Microscopic picture of the blister showing the external materials.

6.4   Preventive measures against voids, blisters, and delaminations

Voids, blisters, and delaminations are local imperfections and damages, which may occur during the production of plastic pipes or under service conditions. Foreign particles and entrapped air or gases inside plastic material may cause voids inside a thermoplastic pipe. In addition to this, in composite materials such as glass-fiber-reinforced pipes, delaminations may be causes during fabrication of the pipe. In glass-fiber-reinforced pipes, osmosis may lead to blisters and delaminations. Hence, measures to prevent these local failures should be related to the above-mentioned causes. Quality control of aggregates used in production of thermoplastic materials should assure that no foreign particles are present during the extrusion of the pipe.

In composite pipes, the composition of the constituting materials and the pipe wall structure should be such that the material incompatibility, entrapped air, and local imperfections are eliminated. In service conditions, in high alkaline environment and under concentration gradients of chemicals the osmotic behavior of the composite wall should be improved by appropriate material choice and material composition. Delamination of the pipe liners from the host pipe wall should avoided by the appropriate chose of liners, resins, placement of liners, and the treatment of the liner system specially at the critical sections along the piping system.

Improper storage and transport can cause local damages in pipe. In composite pipes, improper handling can cause local cracks and delaminations. The case of local cracks has been discussed in Chapter 3. Care should be taken in the handling of pipes during the storage period and their transport and installation.

7
Fatigue, corrosion, and wear

7.1 Introduction

In this chapter, three causes of material deterioration are discussed; these aspects are related to various phenomena, which are caused by different mechanisms. However, the common feature of all three is the reduction of material qualities, which is partly responsible for failures and shortening of pipe service life.

7.2 Fatigue

Fatigue is the loss of strength or other mechanical properties as the result of stressing over a period of time. Fatigue can be divided into several categories; these include: (1) cyclic fatigue and (2) static fatigue or creep rupture (Table 7.1).

7.2.1 Cyclic fatigue

Cyclic fatigue can be defined as the stress, strain, and deformation induced in a material by cyclic loading. *Cyclic fatigue life* is the number of loading cycles which produce

Table 7.1 Types of fatigue in polymer materials and plastic pipes.

	Fatigue									
Material	Thermoplastic pipe		Composite pipe				Multi-layer pipe			
Extent	Pipe	Connection	Fiber	Matrix	Fiber-matrix interface	Whole section	Layers	Interface	Whole section	Connection

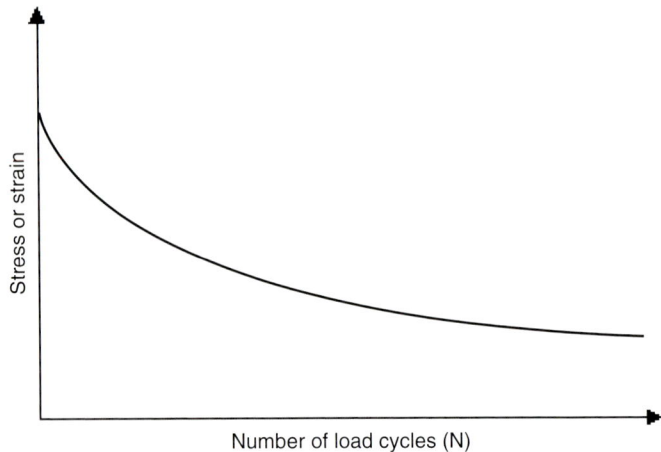

Fig. 7.1. Typical cyclic fatigue behavior of polymers.

a rupture or breakage in the material. Fig. 7.1 shows the typical cyclic fatigue response in a polymeric material. Polymer fibrils subjected to fatigue stress are oriented according to the stress direction. Under normal stress action, the polymer fibrils are oriented along the normal stress. In plastics under shear action, the polymer fibrils undergo lateral shear deformation. In the latter case, shear bands are produced. Each of these two actions may lead to crazing.

Cyclic fatigue may also occur in the pipes under dynamic service conditions. For example, submerged pipes under hydrodynamic forces may experience cyclic-type fatigue. Also, exposed pipes with high velocity of fluid flow may undergo vibratory motions, which may cause fatigue damage in the pipe and its connections. The flutter type instability may also occur in the pipes with high velocity fluid motion.

7.2.2 *Static fatigue or creep rupture*

The fatigue failure may also occur under static conditions provided that the material is subjected to loads for sufficient period of time. This phenomenon is referred to as *static fatigue* or *creep rupture*. The long-term internal pressure tests on the plastic pipe samples at elevated temperatures and the appropriate extrapolation of data (see Chapter 1) are used to produce the creep rupture curves for plastic pipes materials. The curves are represented as hoop stress in the pipe plotted as the time to failure. The curves are normally plotted in double logarithmic scales. In the failure stress versus time diagrams for polyolefin based thermoplastic pipes at elevated temperatures, a three stage behavior of the material can be conceived; these include: (1) the ductile behavior at certain time periods, (2) a brittle behavior at longer times, and (3) a transition zone between the two above-mentioned modes of failure. At transitional zone, the mode of failure changes in character. The failure stresses related to the time periods below this zone is ductile rupture, while those beyond this zone are brittle fracture. The transition zone signifies the qualitative change of the material behavior in the molecular and micro scales. The transition from ductile to brittle behavior of polymers is accelerated by the following factors: increased

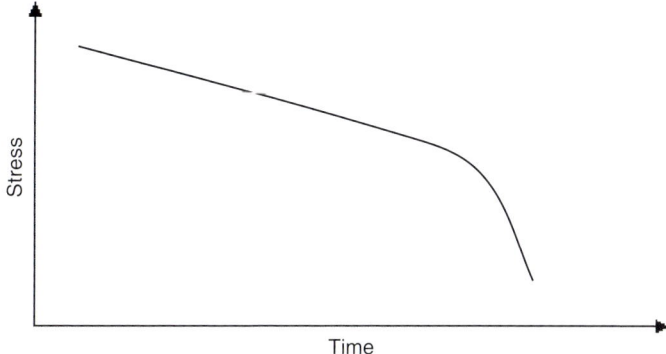

Fig. 7.2. Schematics of creep rupture in plastic pipes.

Table 7.2 Types of corrosion.

	Corrosion/stress corrosion											
Specification	Through thickness		External		Internal		Extent		Crazes/cracks		Color change	
Extent	Whole perimeter	Part of section	Whole perimeter	Part of the surface	Whole perimeter	Part of the surface	Local	Global	Inside	Outside	Inside	Outside

temperature, cyclic loading, stress concentrations, contact with aggressive fluids, and multi-axial loading. Fig. 7.2 shows schematics of the static fatigue in plastics.

7.3 Corrosion

Corrosion is the deterioration of a material as a result of reaction with its environment, especially with oxygen. Although the term is usually applied to metals, all materials, including ceramics, plastics, rubber, and wood deteriorate at the surface to some extent when they are exposed to certain combinations of liquids and/or gases (Table 7.2).

Plastics may react with some external environmental agents. The reaction may take place at the surface or may occur internally due to diffusion of external agents into the plastic body. In the case of composite plastic pipes such as glass-reinforced plastic pipes, the action of external chemical agents may cause corrosion of some components such as the glass fibers. An example of such corrosion is the phenomenon of strain corrosion, which may occur in glass-fiber-reinforced plastics under the influence of acidic environment.

The product of corrosion is cracking and change of physical and mechanical properties of plastic pipes and pipe liners.

Degradation of plastics as the result of corrosion may be caused by one or combination of the following basic sources:

(1) Absorption of some environmental agents into the plastic. The result of this intrusion may be swelling, change of weight, and color. Some aggressive environmental agents may react with the polymer chains. The result of this action is softening and distortion of polymer.

(2) Oxidation of the resinous molecule can occur in the atmosphere or other oxidizing conditions. This often results in hardening (embrittlement) and cracking of the plastic including slow crack growth.

(3) Continued polymerization of the resin can also occur with certain resinous components resulting in hardening, shrinkage, and cracking of the material.

(4) Microbial corrosion, or bacterial corrosion, is a corrosion caused or promoted by microorganisms. It can apply to both metals and non-metallic materials.

The important aspect of the corrosion mechanism is that degradation is not a surface effect like metallic corrosion, but occurs internally in a plastic. A plastic material may absorb an aggressive agent. The environmental stress cracking in plastics is also sometimes referred to as the *stress corrosion*.

7.4   Wear

Wear is a relatively general term implying the limits of material resistance or the loss of surface qualities. Wear can be an indication of the ageing of the material and end of life situation. It may also be defined as the loss of material from the surface due to abrasion (Table 7.3).

Table 7.3 Types of abrasion and wear in plastic pipes.

	Abrasion/wear											
Specification		Medium		External		Internal		Extent		Pealing		Color change
Extent	Fluid	Solid	Whole perimeter	Part of the surface	Whole perimeter	Part of the surface	Local	Global	Local	Global	Local	Global

7.5 Assessment of fatigue, corrosion, and wear

This section contains a basic set of guidelines for failure investigation of plastic pipes as related to the fatigue, corrosions, and wear. Tables 7.4–7.6, in respective fashion, relate various types of fatigue, corrosions, and wear in plastic pipes with the possible causes including mechanical, thermal, chemical, biological, and time factors. These tables are

Table 7.4 Assessment of fatigue.

	Material	Thermoplastic pipe		Composite pipe			Multi-layer pipe				
	Extent	Pipe	Connection	Fiber	Matrix	Fiber-matrix interface	Whole section	Layers	Interface	Whole section	Connection
Influence											
Material	+	+	+	+	+	+	+	+	+	+	+
Mechanical	+	+	+	+	+	+	+	+	+	+	+
Internal pressure	+	+	+	+	+	+	+	+	+	+	+
External pressure											
Axial tension	+	+	+	+	+	+	+	+	+	+	+
Axial compression											
Bending											
Traffic load	+	+	+	+	+	+	+	+	+	+	+
Settlement											
Uplift											
Production											
Impact											
Vibration	+	+	+	+	+	+	+	+	+	+	+
Fatigue	+	+	+	+	+	+	+	+	+	+	+
Residual stresses											
Other											

(Continued)

Table 7.4 (*Continued*)

	Material	Thermoplastic pipe		Composite pipe			Multi-layer pipe				
	Extent	Pipe	Connection	Fiber	Matrix	Fiber-matrix interface	Whole section	Layers	Interface	Whole section	Connection
Fatigue											
Thermal											
High temperature inside											
High temperature outside											
UV radiation											
Fire											
Frost	+	+	+	+	+	+	+	+	+	+	+
Other											
Chemical											
Water											
Oxygen											
Acids											
Alkalis											
Solvents											
Oil											
Benzene											
Other											
Service conditions											
Abrasion											
Interventions	+	+	+	+	+	+	+	+	+	+	+
Other											
Biological											
Microbes											
Animals											
Other											
Ageing factors	+	+	+	+	+	+	+	+	+	+	+
Long-term effects	+	+	+	+	+	+	+	+	+	+	+

Table 7.5 Assessment of corrosion.

Specification	Through thickness		External		Internal		Extent		Crazes/cracks		Color change	
Corrosion/stress corrosion												
Extent	Whole perimeter	Part of section	Whole perimeter	Part of the surface	Whole perimeter	Part of the surface	Local	Global	Inside	Outside	Inside	Outside
Influence												
Material	+	+	+	+	+	+	+	+	+	+	+	+
Mechanical	+	+	+	+	+	+	+	+	+	+	+	+
Internal pressure												
External pressure												
Axial tension												
Axial compression												
Bending												
Traffic load												
Settlement												
Uplift												
Production												
Impact												
Vibration												
Fatigue												
Residual stresses												
Other												
Thermal												
High temperature inside												
High temperature outside												

(*Continued*)

Table 7.5　(*Continued*)

	Through thickness		External		Internal		Extent		Crazes/cracks		Color change	
Corrosion/stress corrosion												
Specification / Extent	Whole perimeter	Part of section	Whole perimeter	Part of the surface	Whole perimeter	Part of the surface	Local	Global	Inside	Outside	Inside	Outside
UV radiation												
Fire												
Frost												
Other												
Chemical												
Water	+	+	+	+	+	+	+	+	+	+	+	+
Oxygen	+	+	+	+	+	+	+	+	+	+	+	+
Acids	+	+	+	+	+	+	+	+	+	+	+	+
Alkalis	+	+	+	+	+	+	+	+	+	+	+	+
Solvents	+	+	+	+	+	+	+	+	+	+	+	+
Oil	+	+	+	+	+	+	+	+	+	+	+	+
Benzene	+	+	+	+	+	+	+	+	+	+	+	+
Other	+	+	+	+	+	+	+	+	+	+	+	+
Service conditions												
Abrasion												
Interventions	+	+	+	+	+	+	+	+	+	+	+	+
Other												
Biological												
Microbes	+	+	+	+	+	+	+	+	+	+	+	+
Animals												
Other												
Ageing factors												
Long-term effects	+	+	+	+	+	+	+	+	+	+	+	+

Table 7.6 Assessment of abrasion and wear.

Specification	Abrasion/wear											
	Medium		External		Internal		Extent		Pealing		Color change	
Extent	Fluid	Solid	Whole perimeter	Part of the surface	Whole perimeter	Part of the surface	Local	Global	Local	Global	Local	Global
Influence												
Material	+	+	+	+	+	+	+	+	+	+	+	+
Mechanical	+	+	+	+	+	+	+	+	+	+	+	+
Internal pressure	+	+	+	+	+	+	+	+	+	+	+	+
External pressure												
Axial tension												
Axial compression												
Bending												
Traffic load												
Settlement												
Uplift												
Production	+	+	+	+	+	+	+	+	+	+	+	+
Impact												
Vibration												
Fatigue												
Residual stresses												
Other												
Thermal												
High temperature inside												
High temperature outside												
UV radiation												
Fire												
Frost												
Other												
Chemical												
Water	+	+	+	+	+	+	+	+	+	+	+	+
Oxygen	+	+	+	+	+	+	+	+	+	+	+	+
Acids	+	+	+	+	+	+	+	+	+	+	+	+

(*Continued*)

Table 7.6 Assessment of abrasion and wear.

	Abrasion/wear											
Specification	Medium		External		Internal		Extent		Pealing		Color change	
Extent	Fluid	Solid	Whole perimeter	Part of the surface	Whole perimeter	Part of the surface	Local	Global	Local	Global	Local	Global
Alkalis	+	+	+	+	+	+	+	+	+	+	+	+
Solvents	+	+	+	+	+	+	+	+	+	+	+	+
Oil												
Benzene												
Other	+	+	+	+	+	+	+	+	+	+	+	+
Service conditions												
Abrasion												
Interventions	+	+	+	+	+	+	+	+	+	+	+	+
Other	+	+	+	+	+	+	+	+	+	+	+	+
Biological												
Microbes												
Animals												
Other												
Ageing factors												
Long-term effects	+	+	+	+	+	+	+	+	+	+	+	+

presented in compact matrix form. The rows of each matrix designate various possible environmental effects and the columns depict the types of fatigue, corrosion, and wear, which may occur in the pipe. The probability of an environmental factor being responsible for a specific buckling event is designated by the "+" sign in the matrix. The empty elements of the tables are indications that no general correlation may exist between the environmental causes and the failure events. However, in a specific case, certain correlation (or lack of correlation) may be established.

It should be emphasized that the entries in Tables 7.4–7.6, as other assessment tables in this book, are based on the knowledge of the author and the inputs from some other experts and the experience gained from various failure investigations. These assessments should only be regarded as general guidelines on which the individual assessments with case-specific modifications and judgment can be based. These tables can be

Fig. 7.3. Floor heating with thermoplastic piping system (diameter about 17 mm, wall thickness about 2.2 mm).

used for failure assessments; they also can be used as a *knowledge base* for creation of an *expert system* for failure diagnosis of plastic pipes.

7.6   A typical failure case study

In this section, a typical case of failure of floor heating plastic pipe systems is briefly discussed. The floor heating systems consist of pipes, which are bent in several locations to form a flat spiral network in the floor of the building. The pipe system is normally embedded in concrete. The floor heating systems are pressure pipes which are operated in a closed loop fashion. The fluid inside these pipes contains some chemicals, which in some cases may be detrimental for the plastic material (Fig. 7.3).

There are specifications for the installation of floor heating systems. Prescription of the minimum radius of pipe bends as multiples of the pipe diameter is among specified installation parameters. Nevertheless, failures have occurred in the pipe as the result of improper bending, improper connection, and also the influence of the internal aggressive medium and external highly alkaline concrete. Cracking of thermoplastic pipes and inward buckling of multi-layer plastic pipes are two main categories of failure in these pipes. The buckling case of multi-layer pipes was considered in Chapter 4 related to buckling failure. The present section deals with a case of pipe cracking due to mechanical and environmental effects. This case is outlined in Table 7.7 accompanied by a number of photographs.

Table 7.7 Failure case of a floor heating plastic pipe system.

Failure case C-1	Thermoplastic pipes in floor heating system
Pipe material and dimensions	Cross-lined polyethylene, polyethylene with partial polybuthene; (diameter about 17 mm, wall thickness about 2.2 mm)
Description of the system	The piping system for transport of hot water for the floor heating. The piping system is placed in concrete. In some locations, the pipe is bent with relatively small radius of bending
Failed part	Pipe
Observed phenomenon	Cracks, in particular ring cracks in the bent zone
Failure description	Cracking
Environmental conditions	Inside: hot water; outside: concrete with high degree of alkalinity
Time to failure	1-year old
Tests performed	IR spectrometry, tensile tests, determination of the degree of cross-linking
Other investigations	Site investigation
Failure cause(s)	Material and pipe weakness, influence of alkalis, improper installation condition and, in particular, relatively sharp bent radius
Suggested corrective actions	Improvement of material and installation conditions
Photo documentation	See Figs. 7.4 and 7.5

Fig. 7.4. Ring cracks in the bent pipes in the floor heating system.

Fig. 7.5. Detail of the cracks in the floor heating system.

7.7   Measures against fatigue, corrosion, and wear

Also fatigue, corrosion, and wear are discussed under a single chapter here, but mechanisms of these phenomena are not the similar. Hence the measures against them are not necessarily the same. The common feature of these three phenomena lies in the fact that they are all caused by the environmental and service conditions. The cyclic fatigue is more serious in composite pipes compared with the singly layer thermoplastic pipes. The bond between the reinforcing fibers and the constituting matrix has a major influence on the fatigue resistance of the composite material. The phenomenon wear discussed in this chapter mainly refers to the abrasion of the pipe material. To improve the abrasion resistance of composite pipes a protective layer at the inner surface in contact with flowing materials is to be used. The corrosion of the pipes is more severe in glass-fiber-reinforced pipes under aggressive environment such as acids appropriate resistive glass fibers should be used.

8
Clogging of the pipe system

8.1 Causes of clogging

Clogging, as the name implies, is partial or total obstruction and hence the hindrance of the function of pipe system. Plastic pipes have smooth inner walls that promote high flow rates and resistance to the formation of deposits, thus preventing clogging. In spite of this, depending on the application and the features of the piping system, clogging may occur in some plastic pipes. Through clogging, the flow of fluid inside the pipe is reduced or totally stopped and hence the main functional requirement of the pipeline is jeopardized. Moreover, clogging can lead to other side effects with far severe consequences.

Clogging can be caused by the buckling collapse of the pipe, detachment of pipe pieces, and clogging of the pipe by the external solid objects. The washed-in soil can also cause clogging of the pipe. Another cause of clogging is the longitudinal sagging, partial settlements, and upheaval of the pipe. In the case of cement and concrete pipes, collapse of the pipe may occur due to the breakage of the pipe. In metallic pipes, rusting of the pipe may lead to its clogging. In plastic pipes, buckling and large deformations are the main causes of clogging. Another common cause of clogging is the intrusion of tree root tips and growth of plants into the pipe system. Generally speaking, corrugated pipes are more susceptible to clogging than the pipes with smooth inner wall surface.

In irrigation pipes used in the agricultural regions, suspended particles or algae may cause clogging of the pipe, especially at the outlets. To prevent clogging, sand filters are normally used. Oils and some chemicals may also facilitate clogging of the pipes.

Clogging of perforated drainage pipes in tunnels is a phenomenon, which may occur due to the wash-in of the drainage sand through the perforations into the pipe. In some cases, the perforated drainage pipe may be covered by some types of geotextiles. This scheme can prevent the wash-in of the sand particles, but can at the same time reduce the hydraulic efficiency of the drainage pipe. In planning of such pipes, large enough diameter should be used to allow sufficient flow capacity and also possibility of robotic inspection and flushing of the debris.

One of the causes of premature clogging is the wash-outs from the construction work into the piping system. For example, cleaning of the concreting machinery in

Table 8.1 Types of clogging.

Specification	Clogging											
	Medium		Extent		Location		Material		State		Response	
Extent	Fluid	Solid	Local	Global	Pipe	Joint	Pipe	External	Cohesive	non-cohesive	Solvable	Insolvable

tunnels and discharging the cement water into the drainage pipe can cause partial clogging of the piping system.

8.2   Types of clogging

Table 8.1 summarizes the basic types of clogging, which may occur in plastic pipes. The classifications in Table 8.1 include local as well as global clogging, types of fluid in the pipe, and behavior of the pipe system. In the case of cable protection pipes, clogging of the pipe would prevent placement of the cables inside of the pipe. This type of clogging mainly occurs due to transverse and longitudinal buckling of the pipe.

In relined and composite pipes, clogging may occur due to tearing off part of the inner layer and detachment of pieces from the pipe wall. The detached pieces would then washed into the pipe and under some circumstances can totally block the flow of the fluid in the pipe.

8.3   Assessment of clogging

This section contains a basic set of guidelines for failure investigation of plastic pipes as related to the clogging assessment. Table 8.2 relates various types of clogging in plastic pipes with the possible causes including mechanical, thermal, chemical, biological, and time factors. This table is presented in compact matrix form. The rows of the matrix designate various possible environmental effects and the columns depict the types of clogging, which may occur in the pipe. The probability of an environmental factor being responsible for a specific clogging event is designated by the "+" sign in the matrix. The empty elements of the tables are indications that no general correlation may exist between the environmental causes and the failure events. However, in a specific case, certain correlation (or lack of correlation) may be established.

It should be emphasized that the entries in Table 8.2 are based on the knowledge of the author and the inputs from some other experts and the experience gained from various failure investigations. These assessments should only be regarded as general guidelines on which the individual assessments with case-specific modifications and judgment can be based. These tables can be used for failure assessments; they also can be used as a *knowledge base* for creation of an *expert system* for failure diagnosis of plastic pipes.

Table 8.2 Assessment of clogging in plastic pipes.

Specification	Clogging of pipe system											
	Medium		Extent		Location		Material		State		Response	
Extent	Fluid	Solid	Local	Global	Pipe	Joint	Pipe	External	Cohesive	non-cohesive	Solvable	Insolvable
Influence												
Material												
Mechanical	+	+	+	+	+	+	+	+	+	+	+	+
Internal pressure	+	+	+	+	+	+	+	+	+	+	+	+
External pressure	+	+	+	+	+	+	+	+	+	+	+	+
Axial tension												
Axial compression	+	+	+	+	+	+	+	+	+	+	+	+
Bending	+	+	+	+	+	+	+	+	+	+	+	+
Traffic load	+	+	+	+	+	+	+	+	+	+	+	+
Settlement	+	+	+	+	+	+	+	+	+	+	+	+
Uplift	+	+	+	+	+	+	+	+	+	+	+	+
Production	+	+	+	+	+	+	+	+	+	+	+	+
Impact	+	+	+	+	+	+	+	+	+	+	+	+
Vibration												
Fatigue												
Residual stresses												
Other												
Thermal												
High temperature inside	+	+	+	+	+	+	+	+	+	+	+	+
High temperature outside	+	+	+	+	+	+	+	+	+	+	+	+
UV radiation												
Fire	+	+	+	+	+	+	+	+	+	+	+	+
Frost	+	+	+	+	+	+	+	+	+	+	+	+
Other	+	+	+	+	+	+	+	+	+	+	+	+
Chemical												
Water												
Oxygen												
Acids												
Alkalis												
Solvents												
Oil												
Benzene												
Other												

(Continued)

Table 8.2 (*Continued*)

	Clogging of pipe system											
Specification	Medium		Extent		Location		Material		State		Response	
Extent	Fluid	Solid	Local	Global	Pipe	Joint	Pipe	External	Cohesive	Non-cohesive	Solvable	Insolvable
Service conditions	+	+	+	+	+	+	+	+	+	+	+	+
Abrasion												
Interventions												
Other	+	+	+	+	+	+	+	+	+	+	+	+
Biological												
Microbes	+	+	+	+	+	+	+	+	+	+	+	+
Animals	+	+	+	+	+	+	+	+	+	+	+	+
Other												
Ageing factors												
Long-term effects												

8.4 A failure case investigation

The failure cases due to buckling, which may lead to clogging of the pipe system were treated in Chapter 4. In this section another failure case is dealt with, which has some relevance to buckling and large deformation phenomenon. Breakage of pipes from other materials is not treated in this chapter. Table 8.3 summarizes a case of clogging of some of the pipes in a piping network in a chemical installation.

8.5 Preventive measures against clogging

Clogging of the piping systems may occur due to various causes, which have been discussed in this chapter. In plastic pipe systems, three major causes are large deformations, detachment of pieces from the pipe wall, and depositions in the pipe. Hence, the preventive measures should assure that these probabilities are reduced. The case of large deformations of the pipes and pipe liners and the related preventive measures has been discussed in Chapter 4. Detachment of pieces from the inner layer of composite pipes should be prevented by appropriate material design and proper manufacturing procedure. The probability of clogging by depositions should be reduced by periodic pressure washing of the pipe system.

Table 8.3 Case of partial clogging of a sewer polyvinyl chloride (PVC) pipe in a chemical installation.

Failure case O-1	Clogging of sewer thermoplastic pipe in a chemical installation
Pipe material and dimensions	PVC pipe (diameter, 160 mm; wall thickness, 3.2 mm).
Description of the system	PVC non-pressure piping system in a chemical laboratory, embedded in concrete under a foundation plate, which was supported on piles and was on the top of the concrete-embedded PVC pipe. The pipe embedding concrete rested on the ground at the bottom and both sides; at the top, it was connected to the mat foundation with vertical strips 4 m apart. The piping system was spread throughout the base foundation under the floor of the factory that produced medical fluids. The diameter of the pipes varied from 160 to 100 mm. In addition to chemical fluids, the piping system was also used to drain the products of the toilets in the second floor of this building.
Failed part	Several collapse pipes in the piping network
Observed phenomenon	Large deformations to the stage of folding in many places up to total blockage (clogging). The large deformations occurred not only in the zones which were directly exposed to fluids, but also in the unused zones in which no flow would occur.
Failure description	Highly deformed pipe specially at the ends, irregular deformations documented by video.
Environmental conditions	Relatively high temperature (probably up to about 80 °C), chemical agents (about 2% concentrations: solvents, distilled water, demineralized water).
Time to failure	9 years in service
Tests performed	Comparison tests with similar pipe, torsion pendulum tests
Other investigations	Site investigations, examination of video records
Failure cause(s)	High temperature, chemical actions superposed on inappropriate installation and insufficient stiffness.
Suggested corrective actions	Better choice of material and system; better care in particular installation.
Photo documentation	See Figs. 8.1 and 8.2

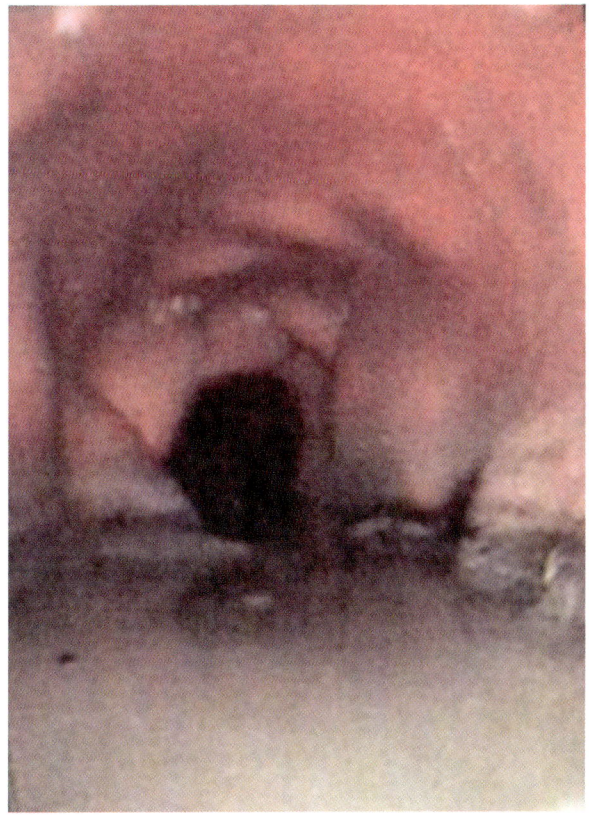

Fig. 8.1. Strong deformation of the polyvinyl chloride (PVC) pipe documented by the video.

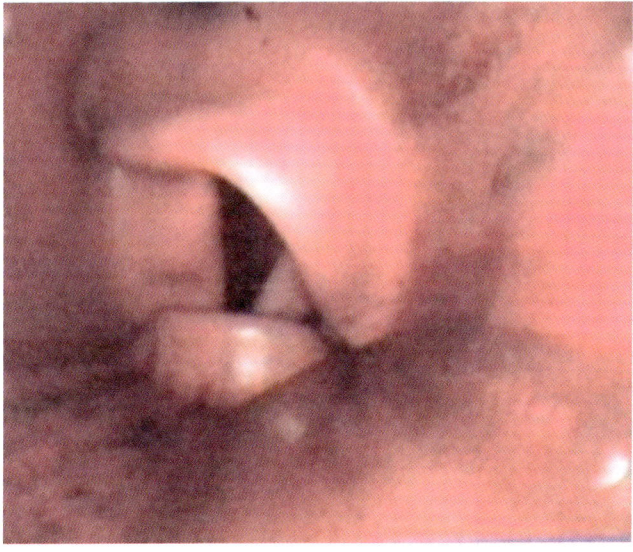

Fig. 8.2. Another video picture showing the large deformation and distortion of the pipe.

9

The knowledge base of an expert system for failure diagnosis

9.1 Introduction

Expert systems are predictive systems utilizing available knowledge and personal experiences to make statements about certain phenomenon or decisions about certain action. The number of developed expert systems in various areas of science and engineering is immense and the survey of the literature in an extremely difficult task and in any case for the present contribution does not serve any purpose. An expert system utilizes a knowledge base as the main "brain" to address a question or an inquiry.

There are a number of expert systems that deal with planning of certain types of piping systems. An expert system dealing with the failure investigation of pipelines, including plastics and composite pipelines, seems however to be lacking. One of the objectives of an envisaged expert system dealing with the failure analysis of pipes would be the clarification and elucidation of the causes of failure. Other objectives would include diagnostic analysis of an existing or an ageing pipeline. Through such investigation, the existing or the potential damages should be identified and the sources of potential failures should be identified. This information would then provide a guideline for the reliability analysis and for the measures for rehabilitation or retrofitting of the system.

All expert systems are founded on a so-called *knowledge base*, which contains the available knowledge in the related field. The knowledge base forms the core of the automated expert system programs. The knowledge base of expert systems for plastic pipes contains the experience of many experts in the area of piping systems and their failure investigation. This coordinated knowledge becomes available for the user at the user level in which the failure investigation of specific pipeline is carried out.

An expert system called EFAP (Expert System for Failure Analysis of Pipes), developed by the author, addresses the above-mentioned objectives. It deals with the explanation of failure causes and aims at tracing these causes to the environmental factors and the life stages of a specific pipeline. This chapter outlines the essentials of this expert system. In this chapter, the *core program* of this system constituting the *knowledge base* is outlined. Fundamentals of the knowledge base of EFAP have been outlined in the previous chapters. The assessment tables presented in each chapter can be cast in a unified framework and can be linked to other modules to build a unified knowledge base structure.

Goal of the expert system EFAP is to provide a first basis for the assessment of the failure, which has occurred in a plastic pipeline. The structure of the expert system is very simple and easy to follow. This expert system is based on a combination of *forward reasoning* and *backward reasoning*. In forward reasoning mode, one can arrive at the failure cause by proceeding from the failure event to failure mode and from there to failure analysis and assessment. In the backward reasoning mode, the failure modes are traced back to the cause of the failure. In each of these two modes, the expert asks questions, which are to be sequentially answered by the user. Through this systematic process, the failure analysis and failure assessment matrices are completed.

This expert system may be considered a *learning system*, in the sense that the user may include the personal experiences and know-how, and hence to enhance the modules of the original system. From this point of view, the present expert system may be regarded as knowledge-based expert system. It is a user-oriented means of assessing a failure event and to take subsequent actions. This manual can also be used as a guideline for investigating the *situation* of existing pipelines. In this capacity, it may be used as a means of *life assessment*.

The expert system EFAP helps to find the causes of the potential or the occurred failure in a plastic pipeline and to make an assessment of its reliability. It is a guideline, which may be used for an ad hoc clarification of the possible failure causes in a plastic pipeline as well as means of systematic failure and reliability assessment. It goes further to make suggestions about the necessary tests and the rehabilitation measures. The core element of the expert system is this manual, which is connected to a number of satellite elements.

9.2   A short description of the expert system EFAP

9.2.1   Structure of the expert system

The expert system EFAP is based on a communication between the user and the software. It is indirectly a technical dialog in the sense that the responses of user are taken into account in the assessments made by the program and that the user can affect this assessment through the new inputs. The expert system has two levels: one consisting of the core program and the other one constituting a level at which the active interaction with user takes place. The core program is the knowledge base, which contains experiences of various experts in the area of piping systems and their failure analysis. In each specific case, this knowledge shall automatically flow into the user level.

Through further developments, the knowledge base may be refined through a learning process and additional experiences may be added to the expert system. The interaction (user) level contains a series of tables and hints dealing with the pipeline. It also contains systematic questions regarding the specific failure case. Digital coding that is

Table 9.1 Procedure for carrying out a specific failure and reliability assessment.

Input into module	Procedure	Output from module	Input into other module(s)
(1) Failure event	Organize all available data from the specific failure case and systematically store them in the module as the source of information about the specific case	(1) Effects (2) Failure scenarios	(1) Environmental effects matrix (2) Failure mode(s) matrix
(2) Environmental effects	(1) Go through the matrix vertically (2) Answer the questions by finding your relevant environmental effect (3) Make a check-mark in the last column of the matrix	Plausible environmental effects responsible for failure	Failure analysis matrix
(3) Failure modes	(1) List the observed failure modes and their specifications (2) Choose the appropriate failure mode table. (3) Choose the column of the table, which corresponds to the observed specifications. (4) In case several modes are present, repeat the procedure	Failure mode(s) and relevant specifications	Failure analysis matrix
(4) Failure analysis	(1) Enter the verified environmental effect (from module environmental effect) in the designated column	Environmental effects potentially responsible for the occurred failure	Failure assessment matrix

	Action	Input	Output
	(2) Enter input from failure mode(s) matrix (matrices) into related column(s) (3) Compare the crosschecked items with related verified environmental effect. In case two crosschecked marks match, enter a crosscheck into the last column of the failure analysis matrix		
(5) Potential failures	Follow the instruction on the table	Potential failure source(s)	Failure assessment matrix.
(6) Tests	Fill out the results of performed tests by check-marking the "test matrix"	Laboratory simulated failure phenomena	Failure assessment matrix
(7) Data bank	Seek the information on similar cases from the data bank matrix	Results from previous cases.	Failure assessment matrix
(8) Failure assessment	(1) Enter output from failure analysis matrix (2) Enter input from "Test matrix" (3) Enter input from "Data bank"	Hypotheses about the failure sources	Further actions
(9) Reliability analysis	(1) Use the history of the pipeline to determine the failure rate of the system (2) Choose a statistical distribution function (3) Use the relations provided or statistical graphs to find the reliability.	Failure assessment	Further actions

1 and zero or equivalently no answer should provide the answer to the questions raised by the expert program. The designated code 1 implies that the answer of the user to the question raised by the expert system is positive. The zero code or the empty space in front of the question shall be considered as a negative answer or a lack of knowledge about the failure event and the failure mode.

The stages in the usage of the program EFAP are as follows:

(1) Visual assessment of the environmental effects that might have acted on the pipe and might have had influence on its failure. For this purpose, a guiding questioner has been made available. The user needs only to enter a digital coding 1 or zero for the existence and nonexistence of the action, respectively. One may enter this coding for one or several questions.

(2) The expert program contains a number of tables dealing with the details of each individual failure mode. The tables include various types of cracks, excessive deformation, buckling instability, color changes, delamination, dimensional changes, abrasion, corrosion, and functional failure. These tables have various cells, which can be activated by the user through the digital response to the questions raised in the table. Through a closer examination of the failed object, the user should be in a position to select the appropriate observed mode of failure and to enter a digital code (1 or zero/none) as input in the related *cell*. This response will then be transported to the *main (core) program* and will be processed there together with the other inputs from the user.

(3) In the main (core) program, the inputs of the user regarding the failure events will be weighted and will be compared with the available expertise and the knowledge. The results of this processing will be transferred to the *user level*.

9.2.2 Modules of the expert program

As pointed out, the expert system has a modular structure and each module of this system is a component, which is located in a network. The constituting modules of the expert system are:

(1) Assessment of the environmental effects
(2) Failure event
(3) Failure modes
(4) Potential failure modes (Tables 9.2–9.9)
(5) Failure analysis
(6) Failure assessment (Table 9.10)
(7) Reliability assessment (Table 9.1)

The totality of modules and their sub-modules are linked to one another and constitute a logical network. Each module can be reached by activating the keyword corresponding to the desired module or work sheet.

Table 9.2 Potential failure modes of the plastic pipeline due to problems caused by applications.

Application of pipeline	Failure mode											
	Crazing	Cracks	Leakage/burst	Deformation	Buckling	Color change	Dimension change	Blisters/voids	Corrosion	Abrasion/wear	Clogging	Deterioration/ disintegration
Water												
Sewerage												
(1) Meteor water												
(2) Building												
(3) Terrain												
(4) Industrial												
Gas												
Protection												
(1) Cable protection												
(2) Power/signal												
(3) Force cable												
Heat/cooling pipes												
(1) Floor heating												
(2) Heat transfer												
(3) Heat exchanger												
(4) Cooling												
Ventilation												
Solid transport												
Oil pipes												
Medical tubing												

Table 9.3 Potential failure modes due to insufficient planning and dimensioning.

Planning and dimensioning	Failure mode											
	Crazing	Cracks	Leakage/burst	Deformation	Buckling	Color change	Dimension change	Blisters/voids	Corrosion	Abrasion/wear	Clogging	Deterioration/disintegration
Project planning												
Dimensioning												
Observation and quality control												

Table 9.4 Potential failure modes of pipelines due to material problems.

Material	Failure mode											
	Crazing	Cracks	Leakage/Burst	Deformation	Buckling	Color change	Dimension change	Voids/blisters	Corrosion	Abrasion/wear	Clogging	Deterioration/disintegration
Thermoplastics												
Thermosets												
Elastomers (1) Natural rubbers (2) Thermoplastic elastomers												
Short fiber-reinforced plastics												
Filament wound plastics												
Sandwich materials												
Recycled polymers												

Table 9.5 Potential failure modes of plastic pipeline due to production problems.

Production method	Failure mode											
	Crazing	Cracks	Leakage/Burst	Deformation	Buckling	Color change	Dimension change	Voids/blisters	Corrosion	Abrasion/wear	Clogging	Deterioration/disintegration
Extrusion												
(1) Single extrusion												
(2) Co-extrusion												
Injection molding												
Resin transfer molding												
Centrifugal												
Filament wound (FW)												
(1) FW alone												
(2) FW with centrifugal												

Table 9.6 Potential failure modes in plastic pipelines due to storage and transport problems.

Storage and transport	Failure mode											
	Crazing	Cracks	Leakage/Burst	Deformation	Buckling	Color change	Dimension change	Voids/blisters	Corrosion	Abrasion/wear	Clogging	Deterioration/disintegration
Storage												
(1) In the factory												
(2) At the construction site												
Transport												
(1) To the construction site												
(2) Handling within the site												

Table 9.7 Potential failure modes in plastic pipelines due to jointing problems.

Jointing method	Crazing	Cracks	Leakage/burst	Deformation	Buckling	Color change	Dimension change	Voids/blisters	Corrosion	Abrasion/wear	Clogging	Deterioration/ disintegration
Butt weld												
Sleeve heated joint												
Electro-fusion												
Push-in sockets												
Screwed jointing												
Glued joint												
Flange connection												
(1) Socket with rubber profile												
(2) Flange with screws												

Table 9.8 Potential failure modes of plastic pipelines due to installation problems.

Installation	Crazing	Cracks	Leakage/burst	Deformation	Buckling	Color change	Dimension change	Voids/blisters	Corrosion	Abrasion/wear	Clogging	Deterioration/ disintegration
Buried												
(1) In earth												
(2) In concrete												
(3) In another pipe												
Exposed												
(1) On supports												
(2) On the floor												
(3) Suspended												
Submerged												
(1) Buried												
(2) On the seabed												

Table 9.9 Potential failure modes of plastic pipeline under the operation and repair conditions.

Environmental influences	Failure mode											
	Crazing	Cracks	Leakage/burst	Deformation	Buckling	Color change	Dimension change	Voids/blisters	Corrosion	Abrasion/wear	Clogging	Deterioration/ disintegration
Mechanical												
Thermal												
Chemical												
Biological												
External interference												
Natural catastrophes												
Inappropriate service												
Inappropriate maintenance												

Table 9.10 Assessment on the basis of potential failures.

Verified environmental effects	Loading and environmental actions	Application	Engineering	Material	Production	Storage	Transport	Installation	Acceptance	Operation	External interference	Natural
	Mechanical											
	Internal pressure											
	External pressure											
	Axial tension											
	Axial compression											
	Traffic load											
	Settlement											
	Uplift											
	External interference											
	Impact											
	Vibration											
	Fatigue											
	Residual stresses											
	Other											
	Thermal											
	High temperature inside											

(*Continued*)

Table 9.10 (*Continued*)

Verified environmental effects	Loading and environmental actions	Application	Engineering	Material	Production	Storage	Transport	Installation	Acceptance	Operation	External interference	Natural
	High temperature outside											
	UV radiation											
	Fire											
	Frost											
	Other											
	Chemical											
	Water											
	Oxygen											
	Acids											
	Alkalis											
	Solvents											
	Oil											
	Benzene											
	Other											
	Operation											
	Abrasion/wear											
	External interference											
	Other											
	Biological											
	Microbes/bacteria											
	Insects											
	Other											
	Time effects											
	Long-term effects											
	Creep											

9.3   Failure analysis procedure by EFAP

Fig. 9.1 shows the systematic procedure for the failure investigation of a pipe using the expert system EFAP. The failure modes occurred in the pipe are incorporated in a failure analysis matrix. Additionally, the failure analysis matrix embodies other elements from the expert systems, which are embedded in the main program. These elements are the expert opinions including the previous experiences on the failure of pipelines.

The core of the expert program as well as the user level contains detailed tables for each failure mode. For example, a buckling event may include varieties of instability involving local or global buckling in the circumferential or longitudinal directions, and so on. As further example, a longitudinal crack in the pipe may be classified into through cracks or surface cracks, individual cracks or multi-cracks, regular or irregular cracks, and so on. The user should assess the occurrence of a particular mode of failure by a digital coding. A categorical assessment has also been made independently by the experts and is embedded in the core (knowledge base) of the main program.

The input from the user regarding his assessment of the observed failure mode and the environmental effects are compared with the assessment of the pipe experts at the program core and are transferred to the user window. The result of such constructive confrontation and synthetic comparison is a so-called "failure analysis matrix." The rows of the failure analysis matrix indicate the possible environmental effects, while the columns of the matrix indicate the possible failure modes.

The failure analysis matrix is the first product of the system. It compares the expert data available inside the core program with the input from the user. The result of this comparison is a "judgment" regarding the environmental factor(s) responsible for the failure event. This judgment appears in the last column of the failure analysis matrix. The failure analysis and diagnosis of the pipe are automatically carried out based on the assessment tables shown in Chapters 3–8. Using this knowledge base as the foundation, a number of automatic assessment matrices are prepared. One of these assessment matrices traces the environmental effects responsible for the failure to the certain stage of the service life of the pipeline. Another matrix identifies the observed failure mode with the certain stage of the service life. In the automatic user-oriented expert system program, an assertion of certain effect or condition is indicated by the digital code 1, while the lack of such influences is signified by the code zero (0).

9.4   Failure analysis

The product of the failure analysis matrix is one or more hypotheses about the failure causes. The hypothesis (hypotheses) gives (give) a guide as to the type of the environmental influences, which might have been responsible for the occurred failure (Fig. 9.1 and Table 9.11).

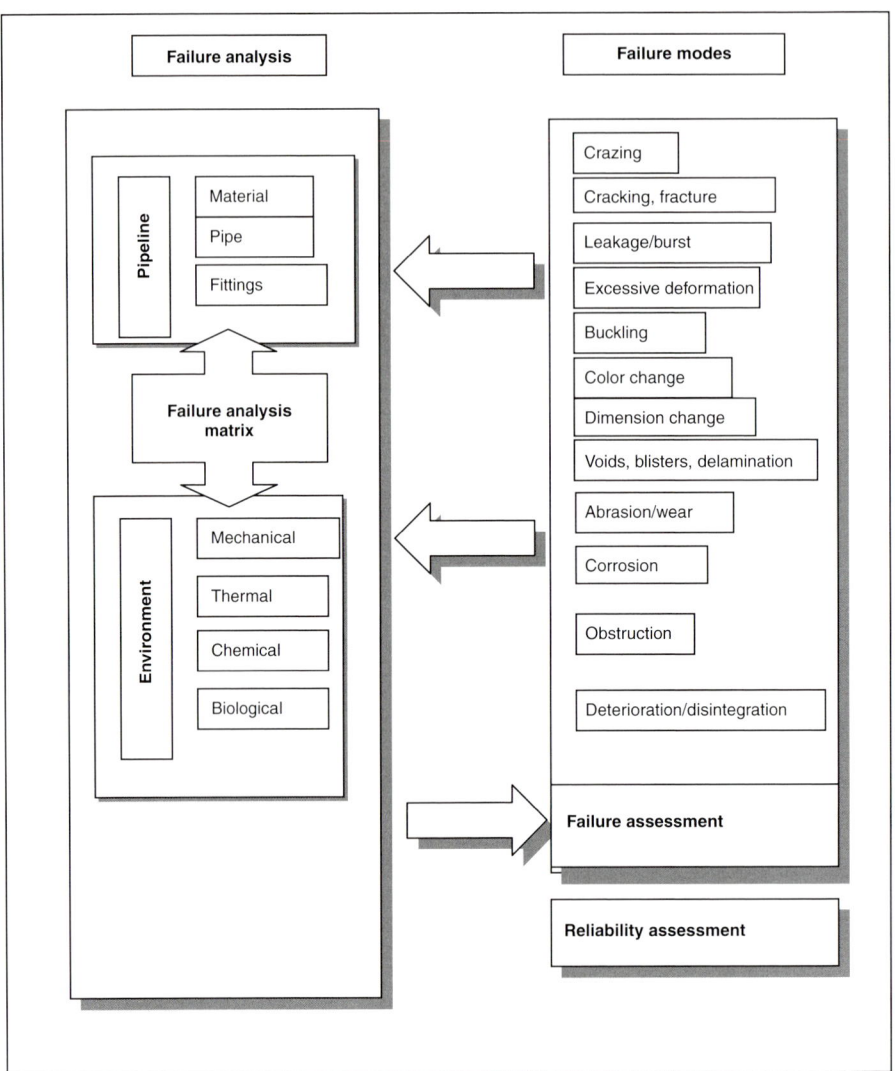

Fig. 9.1. An overview of relation between failure modes and failure analysis modules.

Table 9.11 Failure analysis matrix.

Failure modes and failure causes

Influences	Abbreviation	Observed action	Crazing	Cracks	Deformation	Buckling	Color change	Dimensional change	Blisters/voids	Corrosion/stress corrosion	Abrasion/wear	Obstruction	Pipe break	Failure causes
Material	MAT	+	−	+	−	−	−	−	−	−	−	−	+	**Material**
Mechanical	MEC	+	−	+	−	−	−	−	−	−	−	−	+	**Mechanical**
Internal pressure	IPR	+	−	+	−	−	−	−	−	−	+	−	+	Internal pressure
External pressure	EPR	+	−	−	−	−	−	−	−	−	−	−	+	External pressure
Axial tension	ATE	−	−	−	−	−	−	−	−	−	+	−	+	−
Axial compression	ACO	−	−	−	−	−	−	−	−	−	−	−	+	−
Bending	BEN	+	−	+	−	−	−	−	−	−	+	−	+	Bending
Traffic load	TRA	+	−	+	−	−	−	−	−	−	−	−	+	Traffic load
Settlement	SET	−	−	−	−	−	−	−	−	−	−	−	+	−
Uplift	UPL	−	−	−	−	−	−	−	−	−	−	−	+	−
Production	PRO	−	−	+	−	−	−	−	−	−	−	−	+	−
Impact	IMP	−	−	−	−	−	−	−	−	−	+	−	+	−
Vibration	VIB	−	−	−	−	−	−	−	−	−	+	−	+	−
Fatigue	FAT	−	−	−	−	−	−	−	−	−	−	−	+	−
Residual stresses	RST	−	−	+	−	−	−	−	−	−	−	−	−	−
Other	OTH	−	−	−	−	−	−	−	−	−	−	−	+	−
Thermal	THE	−	−	−	−	−	−	−	−	−	−	−	−	
High temperature inside	HTI	−	−	−	−	−	−	−	−	−	−	−	−	
High temperature outside	HTO	−	−	−	−	−	−	−	−	−	−	−	−	
UV radiation	UVR	−	−	−	−	−	−	−	−	−	−	−	−	
Fire	FIR	−	−	−	−	−	−	−	−	−	−	−	+	−
Frost	FRS	−	−	+	−	−	−	−	−	−	+	−	+	−
Other	OTH	−	−	−	−	−	−	−	−	−	−	−	+	−
Chemical	CHM	−	−	−	−	−	−	−	−	−	−	−	−	
Water	WAT	−	−	−	−	−	−	−	−	−	−	−	−	

(*Continued*)

Table 9.11 (*Continued*)

Influences	Abbreviation	Observed action	Crazing	Cracks	Deformation	Buckling	Color change	Dimensional change	Blisters/voids	Corrosion/ stress corrosion	Abrasion/wear	Obstruction	Pipe break	Failure causes
Oxygen	OXY	−	−	−	−	−	−	−	−	−	−	−	−	−
Acids	ACI	−	−	−	−	−	−	−	−	−	−	−	−	−
Alkalis	ALK	−	−	−	−	−	−	−	−	−	−	−	−	−
Solvents	CLN	−	−	−	−	−	−	−	−	−	−	+	−	−
Oil	OIL	−	−	−	−	−	−	−	−	−	−	−	−	−
Benzene	BNZ	−	−	−	−	−	−	−	−	−	−	−	−	−
Other	OTH	+	−	+	−	−	−	−	−	−	−	−	−	−
Service conditions	SER	−	−	−	−	−	−	−	−	−	−	−	−	−
Abrasion	ABR	−	−	−	−	−	−	−	−	−	−	+	−	−
Interventions	INT	−	−	+	−	−	−	−	−	−	−	+	−	−
Other	OTH	−	−	−	−	−	−	−	−	−	+	+	−	−
Biological	BIO	−	−	−	−	−	−	−	−	−	−	−	−	−
Microbes	MBA	−	−	−	−	−	−	−	−	−	−	+	−	−
Animals	ANI	−	−	−	−	−	−	−	−	−	−	+	−	−
Other	OTH	−	−	−	−	−	−	−	−	−	−	−	−	−
Ageing factors	TIM	−	−	−	−	−	−	−	−	−	−	+	−	−
Long-term effects	LTE	+	−	+	−	−	−	−	−	−	−	−	+	Long-term effects

9.5   Failure assessment

The failure assessment module is the final product of a failure investigation. This module includes the information from the previous experience, from data bank, and from the failure analysis modules. This information is collectively organized into a matrix. Through correlation of the above-mentioned components of information, a final failure assessment is made.

Fig. 9.2 shows the components of the failure assessment module. Tables 9.12 and 9.13 are the working sheets for the failure assessment. These tables should be completed by the failure analyst. The left column of Table 9.12 constitutes the environmental effects; this column is exactly the same as the left column of Table 9.11. The rows of Table 9.12 include the hypotheses from the potential failure mode module, the personal judgment or the know-how from the data bank, and the results of the failure analysis matrix. The last two columns of Table 9.12 are the resulting hypotheses and the suggested actions. The left column of Table 9.13 also constitutes the environmental effects; this column is exactly the same as the left column of Table 9.11. The rows of Table 9.13 include the stages in the life cycle of the pipeline excluding recycling.

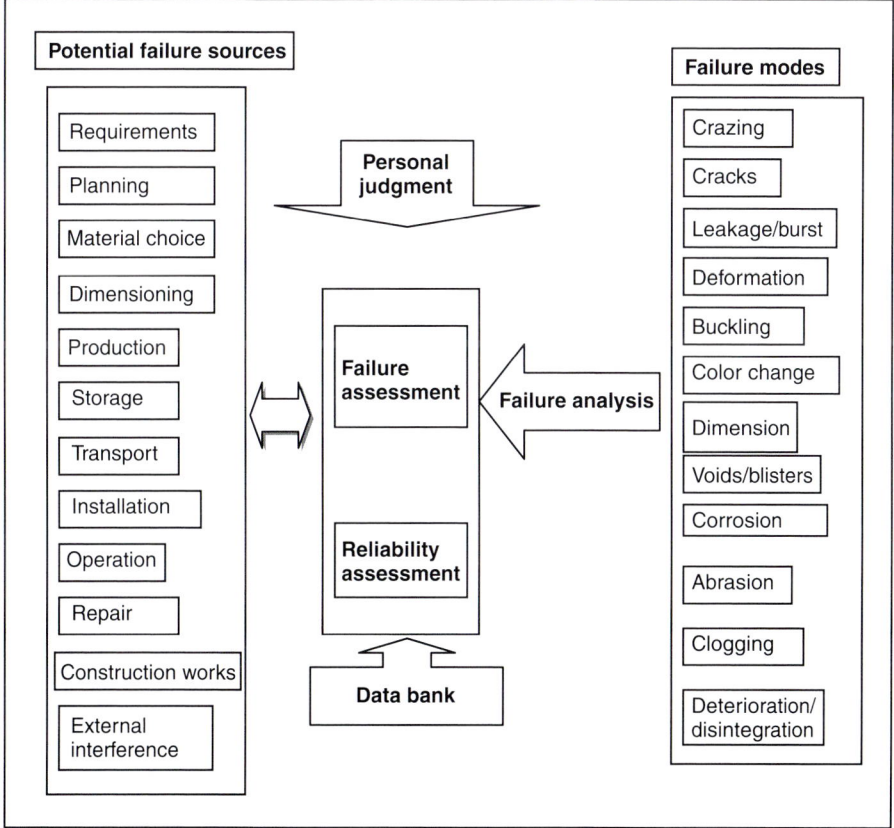

Fig. 9.2. An overview on the failure assessment components.

Table 9.12 Failure assessment matrix – Environmental effects responsible for the failure.

		USE	ENG	MAT	PRO	STO	TRA	INS	ACC	SER	REP	END
Influences	Abbreviation	Usage	Engineering	Material	Production	Storage	Transport	Installation	Acceptance	Service	Repairs	End phase
Material	MAT	+	+	+	+	+	+	+	+	+	+	+
Mechanical	MEC	+	+	+	+	+	+	+	+	+	+	+
Internal pressure	IPR	+	+	+	−	−	−	−	+	+	+	−
External pressure	EPR	+	+	+	−	−	−	−	−	+	+	−
Axial tension	ATE	+	+	+	−	−	+	+	−	+	−	+
Axial compression	ACO	+	+	+	−	−	−	+	−	+	−	+
Bending	BEN	+	+	+	−	+	+	+	−	+	−	+
Traffic load	TRA	+	+	+	−	−	−	+	−	+	+	−
Settlement	SET	+	+	+	−	−	−	−	−	+	−	−
Uplift	UPL	+	+	+	−	−	−	−	−	+	−	+
Production	PRO	+	+	+	+	−	−	−	+	−	+	−
Impact	IMP	+	+	+	−	−	−	+	−	+	+	+
Vibration	VIB	+	+	+	−	−	−	+	−	+	+	+
Fatigue	FAT	+	+	+	−	−	−	−	−	+	+	+
Residual stresses	RST	+	+	+	+	+	−	−	−	−	+	−
Other	OTH	+	+	+	−	+	+	+	−	+	+	−
Thermal	THE	+	+	+	−	−	−	−	−	−	−	−
High temperature inside	HTI	−	−	−	−	−	−	−	−	−	−	−
High temperature outside	HTO	−	−	−	−	−	−	−	−	−	−	−
UV radiation	UVR	−	−	−	−	−	−	−	−	−	−	−
Fire	FIR	−	−	−	−	−	−	−	−	−	−	−
Frost	FRS	+	+	+	−	+	−	+	−	+	+	+
Other	OTH	+	+	+	−	−	−	+	+	+	+	+

(*Continued*)

Table 9.12 (*Continued*)

Influences	Abbreviation	USE Usage	ENG Engineering	MAT Material	PRO Production	STO Storage	TRA Transport	INS Installation	ACC Acceptance	SER Service	REP Repairs	END End phase
Chemical	CHM	+	+	+	−	−	−	−	−	−	+	−
Water	WAT	−	−	−	−	−	−	−	−	−	−	−
Oxygen	OXY	−	−	−	−	−	−	−	−	−	−	−
Acids	ACI	−	−	−	−	−	−	−	−	−	−	−
Alkalis	ALK	−	−	−	−	−	−	−	−	−	−	−
Solvents	CLN	−	−	−	−	−	−	−	−	−	−	−
Oil	OIL	+	+	+	−	+	−	−	−	+	−	−
Benzene	BNZ	−	−	−	−	−	−	−	−	−	−	−
Other	OTH	−	−	−	−	−	−	−	−	−	−	−
Service conditions	SER	+	+	+	−	−	−	−	+	+	−	−
Abrasion	ABR	−	−	−	−	−	−	−	−	−	−	−
Interventions	INT	+	+	+	−	−	−	−	−	+	+	−
Other	OTH	+	+	+	−	−	−	−	−	+	+	+
Biological	BIO	+	+	+	−	−	−	−	−	+	−	+
Microbes	MBA	−	−	−	−	−	−	−	−	−	−	−
Insects	INS	+	+	+	−	+	−	−	−	+	−	−
Other	OTH	+	+	+	−	+	−	−	−	+	−	−
Ageing factors	TIM	−	−	−	−	−	−	−	−	−	−	−
Long-term effects	LTE	+	+	+	−	+	−	−	−	+	−	+

Table 9.13 Failure assessment referring to the stages of the life cycle of the pipeline.

Influences related to the life stage of pipline

Life stage	Abbr.	CRA Crazing	CRA Cracking	TIG Loss of tightness	DEF Excessive deformation	BUC Buckling	COL Color change	SWL Swelling	VBD Voids/blisters/ delamination	COR Corrosion/ stress corrosion	ABR Abrasion/wear	OBS Obstruction
Usage	APP	−	+	+	−	−	−	−	−	−	+	−
Engineering	ENG	−	+	+	−	−	−	−	−	−	+	−
Material	MAT	−	+	−	−	−	−	−	−	−	+	−
Production	PRO	−	+	+	−	−	−	−	−	−	+	−
Storage	STO	−	+	+	−	−	−	−	−	−	+	−
Transport	TRA	−	+	−	−	−	−	−	−	−	+	−
Installation	INS	−	+	−	−	−	−	−	−	−	−	−
Acceptance	ACP	−	+	+	−	−	−	−	−	−	+	−
Service	SER	−	+	+	−	−	−	−	−	−	+	−
Repairs	REP	−	+	+	−	−	−	−	−	−	−	−
End phase	END	−	+	−	−	−	−	−	−	−	−	−

9.6   EFAP – An expert system software

Based on the knowledge based procedure outlined in this chapter a software called EFAP (Expert Failure Analysis Program) was developed. The failure analysis with EFAP identifies the environmental influences including mechanical, physical, thermal, chemical, biological and long-term effects responsible for the failure. Moreover, the failure causes are related to the stages during the service life of the pipeline. Various failure modes including cracking, crazing, large deformation, buckling, leakage and burst, corrosion, surface and body changes, abrasion, delamination, and functional failure are identified. The expert system software has a modular structure including categories of pipelines, potential environmental effects, failure modes, procedure for failure investigation, failure analysis, data bank, knowledge base, atlas of the failure modes, and examples of application. The software has two levels; one is the knowledge base and the other the interactive user window. The product of the expert system software is failure analysis matrix and matrices that relate the failure event to the service life stage of the pipeline. Through further developments, the knowledge base may be refined through a learning process and additional experiences may be added to the expert system.

Glossary of terms and definitions

Abrasion The effects caused by the friction of the flowing fluid and the solids in contact with the pipes on the pipe internal and external surface.

Additive A substance added to another substance to change the properties. Examples of additives in plastics are stabilizers, initiators, and flame retardants.

Ageing The irreversible effects on materials and products of exposure to an environment for an interval of time.

Biological ageing Breakage of molecular chain by micro-organisms.

Blister An imperfection, a rounded elevation of the surface, with boundaries that may be more or less sharply defined. Voids near the internal and external pipe surface. In some cases, large bulging of some layers of a laminated pipe.

Brittle crack A mode of cracking with small strain capacity, characterized by flat, glassy surfaces.

Buckling A mode of instability caused by static and dynamic compressive forces.

Buried pipe A pipe, which is placed at some depth underground and surrounded by earth, sand, or other granular materials.

Burst strength The internal hydrostatic pressure required to cause a pipe or fitting to fail.

Centrifugal process A manufacturing process, in which components of the pipe are thrown against an outside mantle, the pipe is then cured by a prescribed thermal conditioning. The components of glass-reinforced plastic pipes are matrix (especially polyester), filler (especially sand), and short glass fiber.

Chemical ageing Loss of performance and degradation of properties caused by gradual breakdown of polymer molecules into smaller units.

Chemical resistance Perseverance of the molecular structure, physical, and mechanical properties of the polymer against the chemicals (especially solvents).

Co-extrusion A pipe manufacturing process, in which a layered pipe is produced simultaneously through a single extrusion process.

Color change   Change of the original color of the polymeric material due to the action of chemicals, heat, or ultraviolet (UV) radiation.

Corrosion   A mode of material deterioration mainly caused by the action of chemicals.

Crack   A narrow fissure, which may or may not penetrate the entire depth of the material or the thickness of the body; well-defined crack faces identify a crack.

Crazing   *Fine surface cracks* which may occur in plastics, especially amorphous polymers, under stresses smaller than the yield stress. Crazing usually occurs in the direction normal to tensile stress.

Creep   Time-dependent deformation of a material or product under sustained loading.

Degradation   A deleterious change in appearance, chemical structure, physical, and mechanical properties of a plastic.

Delamination   Separation of layers in a laminated body.

Deterioration   A permanent impairment of physical and mechanical properties of plastic.

Diffusion   Penetration of the fluid or gas and damp into the material and its Brownian motion inside the molecular structure.

Dimension change   Macroscopic increase in the volume of a polymeric material as the result of diffusion of low molecules (especially solvents, water) into the surface of the body.

Disintegration   Destruction of the molecular structure of the polymer.

Ductile   A mode of tensile, compression, or shear failure characterized by relatively large plastic strain.

Ductile crack   A crack showing the ductile character.

Durability   Resistance in long time of the material against the environmental effects and in particular resistance against both physical and chemical ageing.

Embrittlement   Embrittling of originally ductile materials (especially polyethylene) in time due to material defects (especially lack of sufficient stabilizer) and due to some actions.

Environment   All factors, which influence the pipeline; these include applied forces, thermal effects, solar radiation, chemicals, biological factors, and the other further actions.

Environmental stress cracking   Cracking caused in a stressed plastic in presence of specific chemicals.

External interference   The intrusion brought about by the factors beyond the intention and expectation of the planners, designers, installers, and the users. The external extrusions may come about unintentionally or intentionally. Examples of unintentional extrusions are repair and changed in the pipeline and the adjacent construction works. An example of intentional intervention is sabotage.

Extrusion   A manufacturing process in which the melted pipe material is being guided through a dye and is then extruded through an opening, and subsequently is being cooled.

Failure   End of functioning of the system according to the requirements. Events in the pipeline, which can hinder its function, change its configuration, jeopardize its integrity, and potentially endanger the environment.

Failure analysis   Investigation of a failure event and clarification of the causes of the pipe failure.

Failure assessment   A judgment about the cause of pipe failure and sources of errors, and also estimation of the pipe life, and the appropriate measures for the life extension.

Failure mode   A feature, which characterizes the failure of pipeline. Examples of these characteristic features are, crazes cracks, color changes, leakage/burst, and delamination.

Failure rate   Probability of failure in the infinitesimal time interval provided that the system has not failed in the time interval (0, t). The start of observation is the time t = 0.

Fatigue   The phenomenon leading to fracture under repeated or fluctuating stresses having a maximum value less than the tensile strength of the material.

Fatigue life   The number of loading cycles which produce a rupture or breakage in the material.

Fiber-reinforced pipe   A polymeric material (called the matrix) which is strengthened by short and long fibers. The fibers are usually glass fibers, but can be other materials. Through reinforcement, the material properties and the pipe performance are normally improved.

Filler   A relatively inert material added to a plastic to modify its strength, stiffness, and other qualities or to lower costs.

Floor heating pipes   An interconnected pipeline, which is placed under the floor and is used for heating of spaces; the pipeline is insolated underneath. The heating takes place by conduction of heat through the pipe and the floor bed and then through convection.

Glass transition temperature   A characteristic index of polymers representing the change of behavior of polymers due to temperature, namely the transition from the glassy state to a rubbery state.

Glassy   A crack face, which is quite flat and resembles a broken glass.

Hoop stress   The stress in the pipe along the circumferential direction.

Intervention   See External interference.

Leakage/burst   Escape of fluid or gas from the pipe, its fittings, and between the pipe and the fitting in the form of diffusion, weeping, leakage, or bursting.

Loading   The environmental effects, which produce stress and deformation in the body.

Long-term hydrostatic strength (LTHS)   The hoop stress in the pipe under internal hydrostatic pressure that when continuously present will cause failure of the pipe.

Maintainability   Probability that, under-defined conditions, the time period required for the sustainability of the system is smaller than a previously defined time interval.

Monomer   A low-molecular-weight substance consisting of molecules capable of reacting with like to unlike molecules to form a polymer.

Obstruction   An event in which part or the whole section of the pipe in a place or in many places may be stuffed with solid material. Obstruction may be in the form of deposition of solid material on the pipe wall or may occur as the result of stuffing of the pipeline by the movement of solid particles or objects.

Osmosis   A pressure difference, which is caused between two sides of a surface or layer through different concentration of a fluid at both sides or two different fluids.

Ovalization   Flattening of the pipe cross-section usually caused by lateral forces.

Oxidation   Breakage of molecular chain by oxidizing agents (oxygen, ozone, some chemicals like nitric acid, sulfuric acid) and hydrolytic action of water. Remedy: antioxidants.

Photo-ageing (photo-degradation)   Degradation of properties due to UV present in sunlight. Remedy: UV stabilizers such as carbon block or titanium dioxide.

Physical ageing   Gradual adverse change in the physical state and order of a plastic due to, for example, migration of additives through the polymer structure of mostly amorphous polymers (especially in PVC).

Pipe stiffness   The ratio of applied load to resulting deformation in pipe section.

Pipeline   A piping system composed of pipe and its fittings.

Plastic   A material that contains as an essential ingredient one or more organic polymers.

Plastic pipeline   A pipeline made of plastic material.

Polymer   A substance consisting of molecules characterized by the repetition of one or more types of monomeric units.

Pressure less pipe   A pipeline, in which the fluid flows under gravity action (especially sewerage pipe).

Pressure pipe   A pipe which is under internal pressure (especially water and gas pipe).

Rapid crack propagation (RCP)   A phenomenon, which is characterized by a fast-moving brittle crack along the pipeline. In pipes, the speed of crack propagation amounts to 250 m/s. The rapidly moving crack has a wavy path and in some places may have branching(s).

Reliability   No failure (structural, functional, etc.) of the system in a defined time period. Reliability is expressed as a probability function.

Relining   A method of pipe rehabilitation, in which another pipe is placed inside the existing pipe.

Residual stresses   The inherent stresses, which are built in a material or the product in the manufacturing stage.

Self-diffusion   Brownian motion of macromolecules in a medium, which is resulted from the chain molecules.

Service life   Time interval between the start of action environmental effects on the system and the failure time at which the system is no longer repairable.

Spalling   Separation of part of the pipe.

Stabilizers   Chemicals used in plastics formulation to help maintain physical and chemical properties during processing and service life. An example of stabilizers is UV stabilizer.

Standard diameter ratio (SDR)   The ratio of the outside pipe diameter to the wall thickness.

Static fatigue (creep rupture)   Failure of a ductile and primarily amorphous polymer in air under-loading occurred after a longer period.

Thermal ageing (thermal degradation)   Breakage of molecular chains due to heat, accelerated by oxidizing agents and also mechanical agents (shearing).

Thermal oxidation   With increase of temperature, the state of polymer undergoes changes, which facilitate the diffusion of oxygen in the molecules. This phenomenon is called *thermal oxidation.*

Through crack   A crack which goes through the whole wall thickness of the pipe.

Uplift   Vertical deformation of a buried pipeline, which is usually caused by the action of the ground water reactive bedding, can also cause uplift.

Voids   Macro- and micro-cavities, which may exist in the material or be caused by the production procedure, service conditions, and chemicals.

Water absorption   One of the indices of polymer response to the water medium or moist environment.

Wear   Failure of a system due to ageing; also, loss of material from the surface due to abrasion.

Weathering   The effects caused by the exposure of the material to the outdoor thermal, radiation, and chemical conditions. Deterioration of the material due to some environmental weathering factors.

Bibliography

1. Ageing pipelines, optimizing the management and operation: low pressure – high pressure, *ImechE Conference Transactions 1999–8*, The Institution of Mechanical Engineers, 1999.
2. Altmeyer, H. (Hsg.), *Kunststoffrohr Handbuch*, 2. Auflage, Vulkan Verlag, 1984.
3. Billmeyer, F.W., *Textbook of Polymer Science*, Wiley-Interscience, 1971.
4. Birolini, A., *Qualität und Zuverlässigkeit technischer Systeme*, Springer-Verlag, 1991.
5. Bocek, M. und Wolf, F., *Ein Verfahren zur Langzeitextrapolation von Zeitstanddaten*, Kernforschungszentrum Karlsruhe, 1983.
6. Brown, T.L. und LeMay, H.E., *Chemie, Ein Lehrbuch für alle Naturwissenschaftler*, VCH, 1988.
7. Bundesamt für Umwelt und Landschaft, Tankanlagen: Richtlinie für Rohrleitungen, SR 814.202, 2003.
8. Bundesgesetz über Rohrleitungsanlagen zur Beförderung flüssiger oder gasförmiger Brenn- oder Treibstoffe, 746.1, Oktober 2000.
9. Cardon, A.H. and Verchery, G. (Editors), *Durability of Polymer Based Composite Systems for Structural Applications*, Elsevier Applied Science, London.
10. Chudnovsky, A., Baron, D. und Shulkin, Y., Lifetime, toughness and reliability of engineering thermoplastics, *Ageing of Material and Methods for the Assessment of Lifetimes of Engineering Plant CAPE '97*, A.A. Balkema, pp. 297–307, 1997.
11. Ehrenstein, G.W. und Pongratz, S., *Thermische Einsatzgrenzen von Kunststoffen in Verarbeitung und Anwendung*, Springer-Verlag, Düsseldorf, 2000.
12. Ehrenstein, G.W., *Kunststoff-Schadensanalyse, Methoden und Verfahren*, Karl Hanser Verlag, München, 1992.
13. European Standard, EN705, Plastics piping systems – Glass-reinforced thermosetting plastics (GRP) pipes and fittings – Methods for regression analysis and their use, 1994 (AC 1995).
14. Farshad, M. and Necola, A., Effect of aqueous environment on the long-term behavior of glass fiber-reinforced plastics pipes in, *Polymer Testing*, Vol. 23, No. 2, pp. 163–167, 2004.
15. Farshad, M. and Necola, A., Strain corrosion of glass fiber-reinforced plastics pipes, *Polymer Testing*, Vol. 23, No. 5, pp. 517–521, 2004.

16. Farshad, M., Determination of the long-term hydrostatic strength of multilayer pipes, *Polymer Testing*, Vol. 24, pp. 1041–1048, 2005.

17. Farshad, M., EFAP – An expert system for failure analysis of plastics pipes, *Plastics Pipes XII,* Milano, 2004.

18. Farshad, M., ADAP – A software for automated design and structural analysis of plastics pipelines, *Plastics Pipes XII,* Milano, 2004.

19. Farshad, M., ADAP: A new software for automated design and structural analysis of plastic pipelines, *Journal of Thermoplastic Composite Materials*, Vol. 16, pp. 171–181, 2003.

20. Farshad, M., ADAP: Neues Rechenprogramm für die Dimensionierung und statische Analyse von Rohrleitungen, *gwa*, Nr. 4, S. 241–249, 2002.

21. Farshad, M., New automated long-term extrapolation method for plastics pipes under hydrostatic pressure, to appear in: *Journal of Thermoplastic Composite Materials*, 2006.

22. Farshad, M., Rohrberechnungsprogramm-Neue Version des ADAP, *gwa*, 1, pp. 899–911, 2004.

23. Farshad, M., Bianchi, S. and Löwe, Ch., Long-term extrapolation of properties of polymeric waterproofing membranes, *Materials and Structures*, Vol. 38, pp. 557–560, 2005.

24. Farshad, M., Sicherheit von Rohrleitungen aus ingenieurmässigen Betrachtungen, gwa, 1, pp. 31–47, 2006.

25. Gas Pipeline Incidents, *5th Report of the European Pipeline Incident Data Group*, EGIG-Group, Document No. EGIG 02.R.0058, 2001.

26. Gas- und Wasserversorgungsleitungen in der Schweiz, Kunststoffe-Synthetics, Nr. 4, S. 30–33, 2003.

27. Gillen, K.T. und Clough, R.L., Time-temperature-dose rate superposition: a methodology for extrapolating accelerated radiation aging data to low dose rate conditions, *Polymer Degradation and Stability*, Vol. 24, pp. 137–168, 1989.

28. Glöckle, W.G. und Nonnenmacher, T.F., Fractional relaxation and the time-temperature superposition principle, *Rheologica Acta*, Vol. 33, No. 4, pp. 337–343, 1994.

29. Griesser, L. and Wieland, M., Earthquake detection and safety system for oil pipelines, *Pipeline and Gas Journal*, Dec 2004, pp. 38–40.

30. Halliwell, S.M., *Weathering of polymers*, (*RAPRA Review*, Vol. 5, No. 5), Pergamon Press, Oxford, 1992.

31. Hawkins, W.L., *Polymer Degradation and Stabilization*, Springer-Verlag, Berlin, 1964.

32. Hertzberg, R.W. and Manson, J.A., *Fatigue of Engineering Plastics*, Academic Press, New York, 1980.

33. Institution of Mechanical Engineerings (ImechE), *Ageing Pipelines, IMechE Conference Transaction 1999–8*, Professional Engineering Publishing, London, 1999.

34. International Standard ISO, Plastics piping systems – Glass-reinforced thermosetting plastics (GRP) pipes and fittings – Methods for regression analysis and their use, *ISO 10928*, 1997.

35. International Standard, ISO, Plastics piping and ducting systems – Determination of the long-term hydrostatic strength of thermoplastics materials in pipe form by extrapolation, *ISO 9080*, 2003(E).

36. ISO, Plastic piping and ducting systems – Determination of the long-term hydrostatic strength of thermoplastics materials in pipe form by extrapolation, *Draft International Standard ISO/DIS 9080*, 1998.

37. ISO, Plastic pipes for the conveyance of fluids under pressure – Miner's rule – Calculation method for cumulative damage, *ISO/FDIS 13760*, 1998.

38. Krebs, C. und Avondet, M.A., *Langzeitverhalten von Thermoplasten: Alterungsverhalten und Chemikalienbeständigkeit*, Carl Hanser Verlag, München Wien, 1999.

39. Lewry, A.J. and Crewdson, L.F.E., Approaches to testing the durability of materials used in the construction and maintenance of buildings, *Construction and Building Materials*, Vol. 8, No. 4, pp. 211–222, 1994.

40. Lu, X. und Brown, N., The ductile-brittle transition in a polyethylene copolymer, *Journal of Materials Science*, Vol. 25, pp. 29–34, 1990.

41. Lu, X. und Brown, N., Unification of ductile failure and slow crack growth in an ethylene-octene copolymer, *Journal of Materials Science*, Vol. 26, pp. 612–620, 1991.

42. Lunquist, L., Leterrier, Y., Sunderland, P. and Månson, J-A, E., *Life Cycle Engineering of Plastics*, Elsevier, Amesterdam, 2000.

43. Farshad, M., Two new criteria for the service life prediction of plastics pipes, *Polymer Testing*, Vol. 23, pp. 967–1072, 2004.

44. Matsumoto, D.S., Time-temperature superposition and physical aging in amorphous polymers, *Polymer Engineering and Science*, Vol. 28, No. 20, pp. 1313–1317, 1988.

45. McManus, H.L., Foch, B.J. und Cunningham, R.A. Mechanism-based modeling of long-term degradation, *Journal of Composites Technology and Research*, Vol. 22, No. 3, pp. 146–152, 2000.

46. Ozawa, T., A new method of analyzing thermogravimetric data, *Chemical Society of Japan Bulletin*, Vol. 38, No. 11, pp. 1881–1886, 1965.

47. Piping systems – Structural design of buried pipelines under various conditions of loading. Part 3: Common method (Statische Berechnung von erdverlegten Rohrleitungen unter verschiedenen Belastungsbedingungen. Teil 3: Gemeinsames Verfahren) prEN 1295–3 (CEN/TC 165 N 1006 E), March 2001.

48. Piping systems – Structural design of buried pipelines: Common CEN method (Draft) CEN/TC164 N 1038, CEN/TC 164/165 JWG 1, N 174a; 1999-05-20, Annex to prEN 1295–2, Preliminary Draft, March 1999.

49. prEN ISO 1167: Thermoplastics pipes for the conveyance of fluids – Resistance to internal pressure – Test method, 2003.

50. Scheirs, J., Compositional and failure analysis of polymers – A practical approach, John Wiley & Sons, LTD, New York, 2000.

51. Schiess, M. and Pluess, C., The Swiss approach to assessing the risk of new natural gas pipelines projects, *Proceeding International Conference on Safety, Risk and Reliability – Trends in Engineering*, Malta, March 21–23, 2001.

52. Schweizerische Rohrleitungsverordnung (RLV) 746.11, März 2000.

53. Still, R.H., *Developments in Polymer Degradation-1*, Applied Science publishers, London, 1979.

54. Structural design of buried pipelines under various conditions of loading. Part 1: General requirements (Statische Berechnung von erdverlegten Rohrleitungen

unter verschiedenen Belastungsbedingungen. Teil 1: Generelle Anforderungen) 155 N 1799, prEN 1295-1 (CEN/TC 164/165/JWG 1, N 47.16 E), August 1996.

55. Sullivan, J.L., Creep and physical aging of composites, *Composites Science and Technology*, Vol. 39, pp. 207–232, 1990.

56. Swiss TS Technical Services AG, Verordnung über die Sicherheit von Druckgeräten, Mai 2003.

57. Tomiita, T., Service life prediction system of polymeric materials exposed outdoors, *Construction and Building Materials*, Vol. 8, No. 4, pp. 223–226, 1994.

58. VDI-Gesellschaft Werkstofftechnik, Bauteildchäden, Verschleiss und Verschleissschutz, VDI-Verlag, Düddendorf, 1995.

59. VDI-Richtlinien, Schadensanalyse- Grundlagen, Begriffe und Defintionen, Ablauf einer Schadensanalyse, VDI 3822, 1984.

60. Wright, D.C., Environmental stress cracking of plastics, *Rapra*, 1996.

61. Wypych, G., *Handbook of Material Weathering*, ChemTec Publishing, Canada, 1995.

Index